ECO HOUSE

ECO HOUSE
PRACTICAL IDEAS FOR A GREENER, HEALTHIER DWELLING

Sergi Costa Duran

FIREFLY BOOKS

A Firefly Book

Published by Firefly Books Ltd. 2010

First printing

Publisher Cataloging-in-Publication Data (U.S.)

Costa Duran, Sergi, 1972–
  The ecological house / Sergi Costa.
[160] p. : col. ill., photos. ; cm.
Includes index.
Summary: Practical ideas for a more environmental and healthy home, including architecture, design, structure and materials.
ISBN-13: 978-1-55407-782-3
ISBN-10: 1-55407-782-6
1. House construction. 2. Dwellings – Environmental engineering.
3. Ecology. I. Title.
690.2 dc22   TH4812.C58 2010

Library and Archives Canada Cataloguing in Publication

Costa Duran, Sergi
 The ecological house / Sergi Costa Duran.
ISBN-13: 978-1-55407-782-3
ISBN-10: 1-55407-782-6
1. Ecological houses. 2. Sustainable architecture.
3. Dwellings – Remodeling – Environmental aspects.
4. Architecture, Domestic – Environmental aspects. I. Title.
TH4860.C62 2010   728'.047   C2010-902395-1

Published in the United States by
Firefly Books (U.S.) Inc.
P.O. Box 1338, Ellicott Station
Buffalo, New York 14205

Published in Canada by
Firefly Books Ltd.
66 Leek Crescent
Richmond Hill, Ontario L4B 1H1

Printed in China

Eco House was developed by:
LOFT Publications
Via Laietana, 32, 4º, Of. 92
08003 Barcelona, Spain
Tel.: +34 932 688 088
Fax: +34 932 687 073
loft@loftpublications.com
www.loftpublications.com

Editorial coordination:
Simone K. Schleifer

Assistant editorial coordination:
Aitana Lleonart

Editor:
Sergi Costa Duran

Art director:
Mireia Casanovas Soley

Design and layout coordination:
Claudia Martínez Alonso

Layout:
Ignasi Gracia Blanco

The author would like to thank all architects, designers and manufacturers for their participation not only for the material provided, but for their daily work to promote eco-design (or just plain good design) and plan their in a more respectful way. Special thanks to the Fundació Terra for being such a source of inspiration for many years.

All prices specified in this publication are guidelines for 2009

CONTENTS

A NEW ARCHITECTURE
FOR A NEW WORLD

Climate change is one of the most worrying socio-environmental phenomena
we face now and in the years to come. This phenomenon has a global reach and,
according to scientists, may result in unpredictable consequences, including
adverse weather conditions, alterations in animal migratory patterns and
changes in water policies and regulation. For this reason, modern construction
– one of the factors contributing to global warming – has come to the fore for
its environmental impact not only on the land, but on climate. This is a problem
that also has local consequences, and for which we need strong and compelling
answers. When we now decide to buy a house or renovate, our decision, just as
when we buy a product or use a service, matters more than ever.

Housing is our third skin. It is a space that protects us from the elements, but
that is also a place of comfort and security. It is where we adore and educate
our loved ones. Perhaps that is why the home, apart from being a right, is an
absolute necessity – as important as food.

Decades ago we became aware of the importance of housing in human
evolution. Other attributes such as comfort or health were to its function as a
shelter. Years later the housing bubble burst, which generated a huge volume
of business for builders, municipalities and speculators of all kinds. For this
reason, architecture over the past 50 years – precisely the architecture that has
witnessed the abandonment of the countryside and the growth of the city – has
put productivity before quality and health. This has resulted in a growing number
of environmental poisons in the form of volatile substances, carcinogenic
materials, confined spaces and energy waste. Time has shown, as does the
current financial situation of some countries today, that an economy based on
bricks is "bread today, hunger tomorrow." Obviously, this has had an impact on
land and society with rising unemployment rates in the most affected countries
and with a badly maintained urban panorama in many suburbs in the wake of
property speculation.

Unfortunately, some of the houses where we live today form part of this unfortunate legacy, which has depleted resources and changed the face of the land where we grew up. This process, it should be reiterated, has a large carbon footprint because of the wasted energy and poor durability of the planned infrastructure. In response to this, energy efficiency and sustainable green standards have emerged in developed countries, in particular in the United States and Canada, such as LEED (U.S. Green Building Council, www.usgbc.org), LEED Canada (Canada Green Building Council, www.cagbc.org), BREEAM Canada (www.breeam.org), Cradle to Cradle C2C (www.mbdc.com) and Energy Star (www.energystar.gov). These are all useful and positive standards to create a new building culture.

This book explains this new direction in architecture, so that in a few decades what may now seem extraordinary will be commonplace. It is a method to build or renovate close to what may eventually be known as bioconstruction – a building system with recycled or recyclable low-environmental impact materials, or materials that can be extracted by simple processes at a low cost. Efficiency is the yardstick for this new architectural direction, and the first chapter is exclusively devoted to various cleaner technologies that can be used in the home, and most importantly, how to make better use of them. The journey continues with a chapter on natural materials for the structure, cladding, floors and walls of a building. The final chapter presents us with a room-by-room action plan for a greener house resulting in the conservation of water, energy and resources.

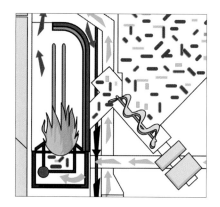
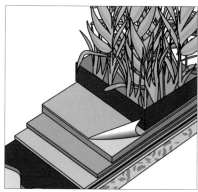
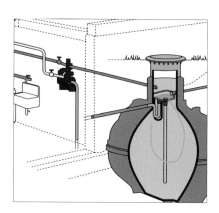

INSTALLATION DIAGRAMS IN THE DESIGN PHASE AND KEY FACTORS FOR A BIOCLIMATIC DESIGN

 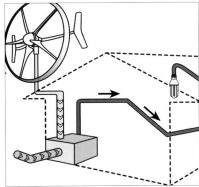 

This series of designs and diagrams helps us better understand the key factors of bioclimatic design for a healthier home, and how a variety of changes can result in significant savings of energy and water consumption. Most of these features should be installed in the design phase of a home, while others can be installed afterward. The efficient work of both an architect and an energy consultant when planning the energy requirements of the house will help optimize these resources.

10 FUNDAMENTAL RULES FOR BIOCLIMATIC DESIGN AND A HEALTHY HOME

1. Main façade facing south. Position eaves depending on the latitude, to provide shade in summer and let sunlight through in winter.

2. Close to deciduous trees, shade in summer.

3. A gallery with large glass expanses on the south side of the house acts as a solar collector.

4. Walls, solid walls and solid materials allow greater thermal inertia, so they accumulate more heat to release later.

5. If the house has a chimney, a thermo-wind self-suction hood, which expels fumes and excess heat and prevents them being sucked back in is recomended.

6. Fit hinged skylights in the roof and adjustable flaps at the bottom of the north face. Skylights light up hallways, bathrooms, attics and other rooms. Since they fold up and can be adjusted, when opened in summer they get rid of the hot air and create cross-ventilation.

7. Use natural insulation in walls and breathable waterproof materials for roofs.

8. Use local building materials when possible.

9. The materials used must be safe in regards to radioactive transmission. Under no circumstances should they emit more than 180 mrad per year, or release radon gas, which is associated with some lung cancers.

10. The home's electrical balance must be in line with the environmental maximum, which ranges from 120 to 300 volts per meter. For that reason synthetic and ferromagnetic materials should not be overused, as they may generate electrostatic charges.

Bioclimatic design of a building

1. Summer
2. Winter

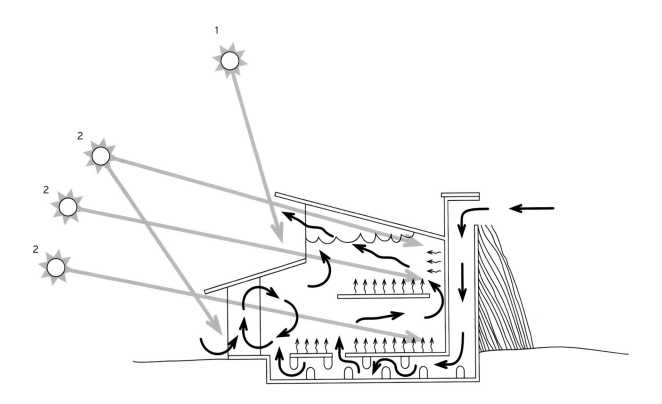

Heat recovery ventilation system

1. Solar thermal panels (optional)
2. Insulation
3. Triple-glazed low-E windows
4. Supply air
5. Extract air
6. Heat recovery ventilation system
7. Ground heat exchanger

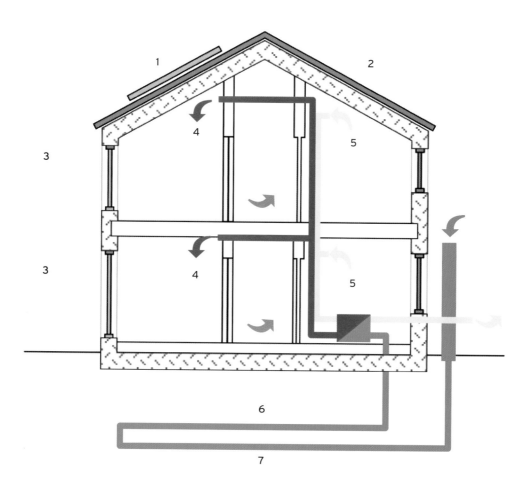

Diagram of roof garden

1. Plant cover
2. Plant substrate
3. Draining membrane
4. Insulation
5. Protective geotextile
6. Roof membrane
7. Structural support

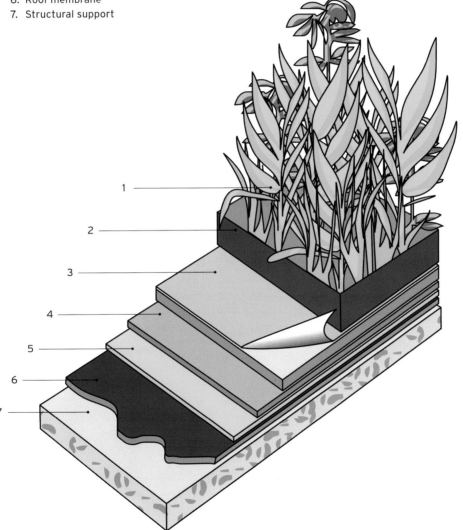

Thermal solar installation

1. Flat collector
2. Underfloor heating
3. Hot water tank
4. Washing machine
5. Kitchen
6. Bathroom

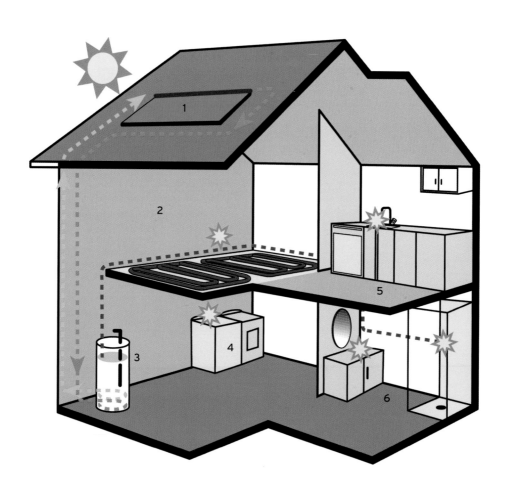

Types of geothermal installations

1. Surface collector: A horizontal loop is installed at a depth of 3–6 feet (1–2 m). It requires a lot of space.
2. Geothermal probe or vertical sensor: It requires much less space in exchange for greater depth, making it ideal for urban areas and apartment blocks.
3. Geopanel: The circuit is built into prefabricated panels that are placed in trenches about 10 feet (3 m) deep. It requires little space and is cheaper.
4. Underground water collector: Underground water can also be used, as long as it not below 50 feet (15 m).
5. Soil temperature
    Summer 57ºF (14ºC)
    Winter 57ºF (14ºC)
6. Ideal temperature
    Summer 73ºF (23ºC)
    Winter 70ºF (21ºC)
7. Outside temperature
    Summer 97ºF (36ºC)
    Winter 36ºF (2ºC)
8. Air convectors
9. Heat pump
10. Underfloor heating
11. Conventional radiator

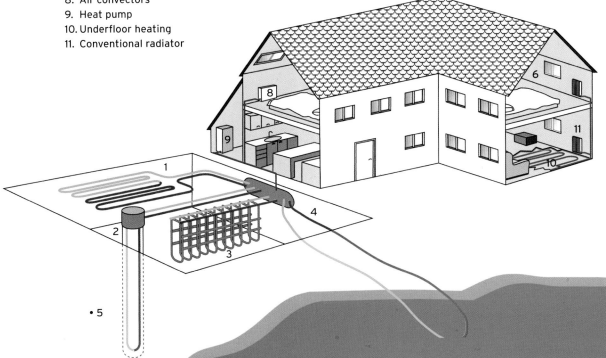

Underfloor heating

1. Ceramics
2. Mortar
3. Binding for pipes
4. Polystyrene insulation
5. Wrought
6. Tube circuit: The water flows through these tubes at a temperature between 93ºF and 115ºF (34ºC and 46ºC)
7. The heat radiated from the floor heats the room to between 64ºF and 72ºF (18ºC and 22ºC)

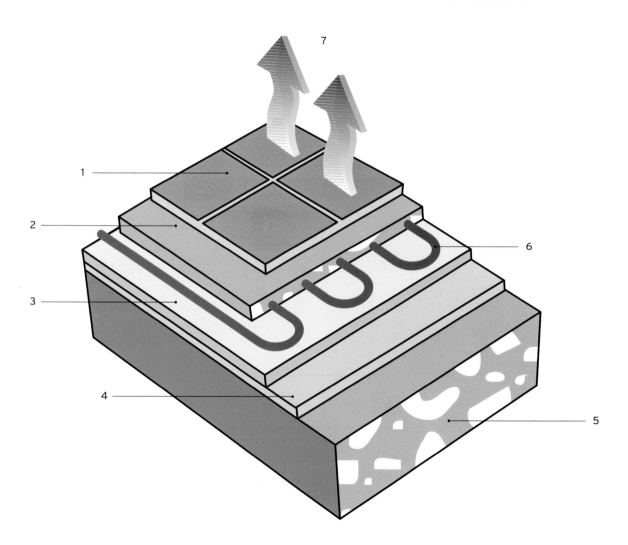

Operational diagram of a pellet stove

1. Fuel tank (pellet)
2. Cochlea for supplying fuel
3. Gear motor
4. Combustion burner
5. Electrical resistance refill
6. Vent pipe
7. Hot air fan
8. Hot air outlet grill
9. Synoptic panel
10. Centrifugal vacuum for smoke

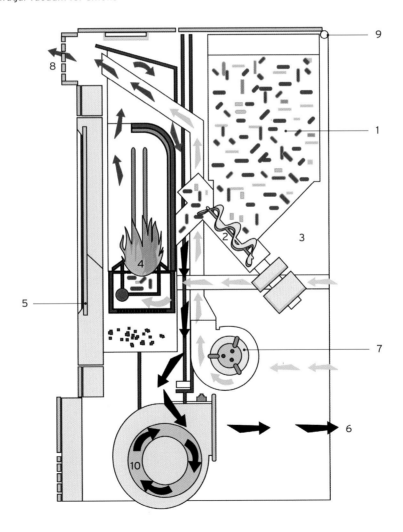

Photovoltaic solar installation

1. Regulator: Protects the battery discharges and overloads, when the installation is in an isolated house
2. Inverter: Converts 12V direct current from panels to 50 Hz and 220V alternating current
3. Meters: Quantifies the electricity in the grid and electricity consumed
4. Protection against external surges that may damage the building
5. Batteries: Only required in isolated houses, which are not connected to the grid

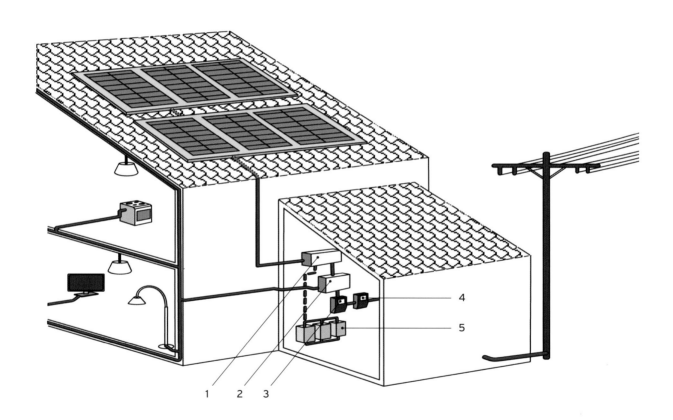

Installation of a mini wind turbine

1. Wind turbine
2. Household power system
3. Transformer
4. Power output

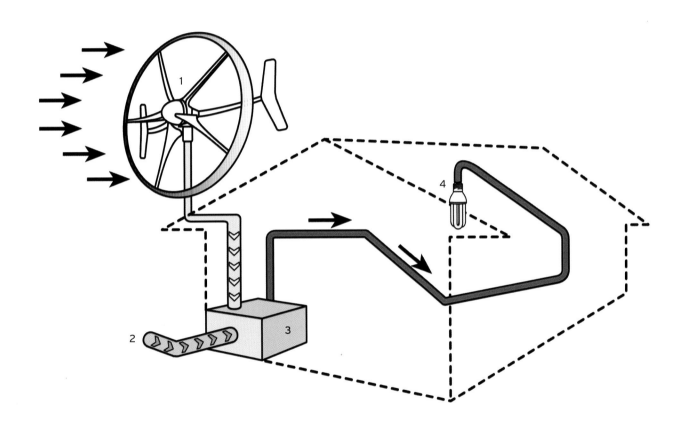

Rainwater collection system

1. Tank
2. PE telescopic cover
3. Filtration equipment
4. Pumping device

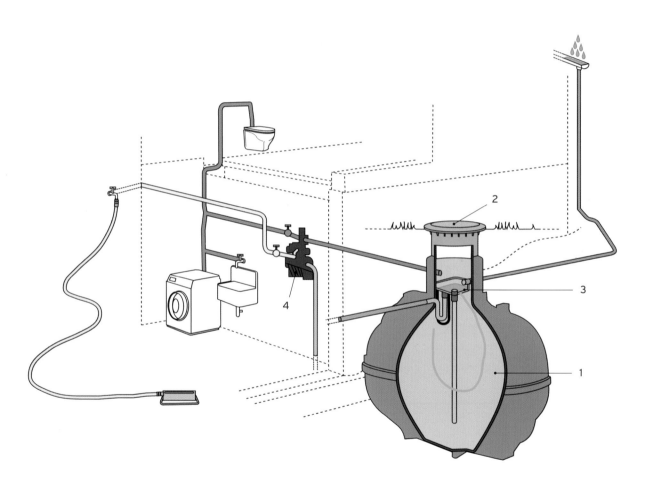

System for purification of graywater

1. Recycled water reused in garden, toilet or to wash the car
2. Process control
3. Overflow to sewer
4. Water from toilet and kitchen to sewer
5. Graywater from bathroom and laundry

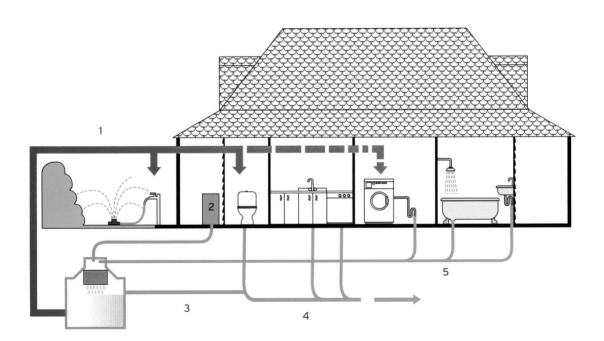

MEASUREMENTS IN THE DESIGN PHASE OF A GREEN HOME

During the design phase the majority of the savings in natural resources, energy and water are determined that will be used in the home during the construction and operation phase. When a house is well designed, last-minute decisions – which arise from poor planning and that will prevent the home from achieving its optimal performance during its long useful life – are avoided.

The more bioclimatic the design, the better, as measures based on the orientation, latitude and climate, which are imposed by the site, should be maximized. These measures, known as passive strategies, where possible should take priority over active strategies (solar photovoltaic, geothermal, radiant floor heating, etc.). Even though active strategies efficiently respond to energy needs, they involve energy consumption, as before they are installed they must be manufactured.

STRUCTURE AND SKIN

The structure of a building and above all its skin is vitally important for saving energy. Establishing a parallel with the human body, some authors have called the home the "third skin": just like clothing, which is the second, it must be able to breathe. And just like actual human skin, which is the first, it must protect but not isolate us at any time from the outside. The healthier the third skin is, the better quality of air we breathe inside the home. Here are a few of the most energy efficient skins.

Adobe

Adobe is formed by a mass of mud (clay, sand and water) sometimes mixed with straw, coconut fiber or dung, molded into bricks and sun-dried for 25 to 30 days. The main mix is 20% clay and 80% sand and water. Its embodied energy is 170 btu/lb. (0.4 MJ/kg). The more embodied energy in a building material, the more energy that has been used during its production.

Adobe is a good acoustic insulator and has high thermal inertia, thus regulating internal temperature. In summer it maintains cool temperatures and in winter, hot. The inclusion of vegetable fibers can attract termites.

If well constructed and maintained, an adobe building can last a hundred years or more. To avoid any cracks, straw, horse hair or dry hay are added to the mass which serve as a framework. For outdoor use, it is recommended in countries with low rainfall.

An alternative to adobe is rammed earth, a large piece of clay compacted inside a wooden mold used to make walls, such as the inner wall of this house in Australia. The mold used for making rammed earth structures, which consists of two parallel panels with a certain distance between them, is the mud-wall.
The most common mold dimensions are about 5 ft. (1.5 m) long, 3 ft. (1 m) tall and 1 $1/2$ ft. (0.5 m) thick. Stabilizers such as straw, manure or lime can be used.

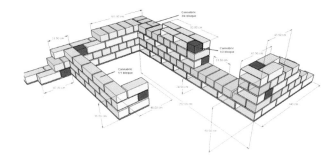

A natural alternative to adobe is Cannabric, a solid block for walls manufactured in Spain since the late 1990s. It consists of plant material, natural binders and minerals and recycled agglomerates without pesticides or herbicides. The resulting ratio is 20 lb (9 kg) of hemp for every square foot.
Cannabric takes advantage of the insulating characteristics of hemp (with a thermal conductivity of 0.008 W/ft.°F or 0.048 W/m°k), much better than that of wood. The mineral components of the block provide mechanical strength, density and high thermal inertia. The blocks are not baked, but air dried for a minimum of 28 days.

Natural stone

Stone is the noble building material par excellence. The most widely used structural materials are granite, gneiss, sandstone, limestone, marble, quartzite and slate. They are used for foundations, walls, façades and as an architectural element. Porous stone is less durable than dense stone. Locally-sourced stone has an embodied energy of 2,500 btu/lb. (5.9 MJ/kg).

Advantages of using stone:
- Durable and easily maintained
- Good soundproofing
- Good thermal inertia, which keeps the temperature stable for walls more than 20 in. (50 cm) wide.
- Good protection from the summer heat
- Warmth

Disadvantages:
- Slower construction and higher labor costs
- Risk of moisture deterioration. In case of frost, the internal water freezes and it expands causing cracks.
- Quarries are overexploited.
- The cutting and polishing stones are energy-intensive, and a lot of waste material is produced.

The use of granite blocks on the façade of this house gives it a monumental air. By way of biomimicry, architects designed the building as a geode, a cavity in the rock in which minerals crystallize as they dissolve in groundwater. Granite is highly radioactive compared with other stones that can be used in curtain walls.

The greenest way to preserve stone is to clean it with a mild sodium or potassium silicate solution and, once dry, apply a $CaCl_2$ solution. These two solutions are called Szerelmey liquid. The calcium silicate solution forms an insoluble layer that protects the stone.

Limestone is a type of stone made of calcium carbonate. It has a low-intensity terrestrial radiation neutralizing property, but it does not react well to lichens. For outdoors, we recommend that the wall is whitewashed to increase its resistance to moisture and to improve its insulation.

For roofs, harder slate such as bitumen or siliceous slate can be used. In this home located in Chile, facing the Pacific Ocean, the architect wanted to use natural materials in keeping with the surroundings of the site.

The embodied energy of the slate is 15,100 btu/lb. (35.1 MJ/kg). It is normally produced from about 85% raw material and 15% resins and polyester. As with other stones this is a renewable natural resource.

Slate can also be used for indoor or outdoor flooring. A durable natural material that is nonslippery and easy to maintain, it can be used in either a square, rectangular, triangular, irregular or customized format.

Marble is carbonated sedimentary rock that has achieved a high degree of crystallization through a process of metamorphism. It is harder, stronger and more durable than limestone. Marble is easy to work with, but it loses its luster if exposed to the elements.

Natural stone can be cut according to the design needs. The exterior walls of this home in Brazil are made of local stone slabs arranged horizontally.

Straw

Straw is a low-embodied-energy building material and it is the most eco-friendly material that can be used to build a home. Thatched buildings date back to the early 20th century in the United States. Straw bales, a surplus product of agriculture and economic development, began to be used as partition walls. Currently this product is popular in Canada, the United States and parts of Europe.

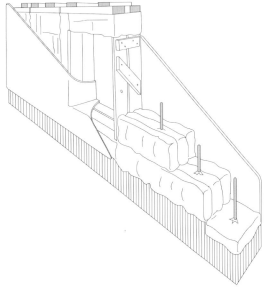

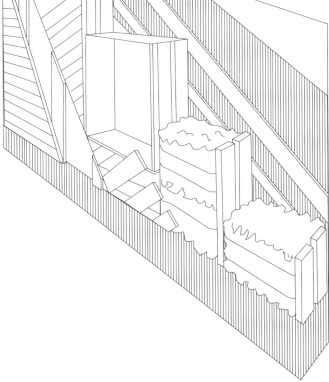

There are two main systems to build a house with straw – the *Nebraska style*, where the straw acts as a load-bearing wall, or in the *post-and-beam* style, in which a system of posts or columns support the weight of the roof. There is also the possibility of a hybrid of both styles. In the first case the size of bundles is significant, which reduces the habitable area.

The bundles of bales are hand-tied with plastic tape. Once positioned, construction is halted so that the material can settle. Given the compaction of the bales, there is not enough air inside the bale for it to burn. Properly maintained, a building of this type can last over 100 years.

To build a house of straw, seek professional help. Even in countries where it is customary it is difficult to find skilled labor. Strictly adhere to regulations on structures and fire protection. It is vital that the straw does not get wet, as it may decay or go moldy. There must not be a space between the bales and the ground as this could become a breeding ground for rodents.

In this house in Switzerland, a lime plaster (a minimum coat of $3/4$ in. [2 cm] is advisable) conceals the straw bale structure, both inside and outside.

The same applies to the northern façade of this home in Canada, which takes advantage of the excellent thermal and acoustic insulation properties of this material in the whitewashed wall. The straw allows the temperature to be naturally controlled inside, even in cold climates.

If part of the straw decays just remove it and replace it with fresh straw (tightly compacted). Experience shows that if part of the wall becomes wet it is usually easy to locate it without affecting the rest of the straw.

Wood

It is *vox pupoli* that the use of wood as a building material,
being of vegetable origin, can be considered as eco-friendly. The
problem is knowing where it comes from and how it has been
exploited, something that is difficult to find out if there are no
green certificates.

Where does most of the fiber used in our homes come from?
Softwood is typically from Europe and the United States. In the
case of hardwood, its origin is Europe (mainly France) and the
U.S. Much of the oak is from Russia, the eucalyptus is almost all
from Latin America. Finally, profiled wood comes from different
parts of the globe, mostly China.

As a general recommendation the more local the wood we use,
the better.

Forests are of vital importance not only as CO_2 sinks, but to
control erosion, promote water infiltration into the soil or to
regulate rainfall.

Currently the label promoted by the Forest Stewardship Council
(FSC), an international organization, is one of the most reliable
certifications to ensure that the wood used in structures,
cladding, flooring and furniture comes from sustainable farms.

Headlands House, in England, is made from a larch and oak structure. The wood is sourced locally, reducing the carbon footprint of the product, as the distance of supply decreases. Other woods that are often used in building are fir, chestnut, cypress, beech and pine.

This is the final appearance of the Headlands house, after the plaster is applied, which will thermally insulate the building. Other options to improve the insulation panels would be the use of coconut fiber, cork or recycled paper.

The durability of a wooden structure depends on the type of wood used and the maintenance. In general, it could exceed 50 years, depending on whether the material has been dried properly or protected from insects. If the cut of the wood is done in the direction of the fibers, a piece of wood is as strong as a similar sized slab of concrete, and has a higher elasticity.

In Mexico, the forest engineer Mario Alberto Tapia Retama reuses pallets to assemble structures for housing that he later insulates with cardboard. These solutions are typical of a subsistence economy, common in some developing countries, but they show that with a little ingenuity and a good sense of a culture of reuse, we can self-construct our home.

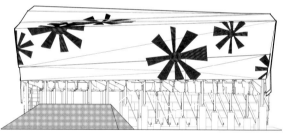

A wooden prefabricated structure can be modeled on the latest cutting technologies. One example is the construction of the prototype Burst 003, by the Australian System Architects, which includes a ribbed structure of laser-cut plywood. Each rib has been previously numbered and cut according to its position. The surface of the outer shell is made with cedar wood.

Bamboo

Bamboo is a woody grass that naturally replenishes itself every seven years and does not require the use of pesticides or fertilizers if cultivated appropriately. Depending on the type, it can grow 3-15 inches (8-40 cm) a day and reach 130 feet (40 m) in three or four months.

It is used in pillars, roofs, ceilings, walls or claddings. If it is used for the structure its maximum strength and elasticity is required (the darker the bamboo, the softer it is). In Latin America and Asia woven cane is used through a type of knotted structure. For cladding, panels can be used.

A bamboo construction also has its environmental problems. To ensure the self sufficiency of this material, native forests are felled to use the land for the cultivation of bamboo, or fertilizers and pesticides are used to increase the yield. For this reason FSC-certified bamboo in the form of veneers or plywood should be used.

It can be used for handrails and stairs, in the picture the use of bamboo combined with spruce (general structure) and redwood (balcony) can be observed.

Brick

Brick is a ceramic material made from clay or a mixture of clay that is molded into blocks that harden in the sun or are baked. It is used in masonry for brick structures, whether they are walls, partitions or façades.

Unlike adobe, which also contains clay, brick is a more processed product, which explains the higher embodied energy: 1,075 btu/lb. (2.5 MJ/kg), six times higher than adobe.

Industrial cooking is carried out in tunnel kilns where the temperature ranges from 1650°F to 1830°F (900°C to 1000°C). For this reason, brick also has less organic waste materials, although part of its composition is natural.

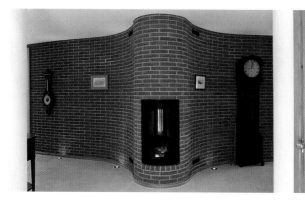

Brick walls have high thermal inertia. In this example the inner wall of this house in the United Kingdom makes use of the heat from the fireplace embedded in the wall. The floor is covered with locally sourced York tiles with underfloor heating powered by a geothermal heat source.

The company Piera Ecocerámic markets the Ecoclinker, an exposed ceramic brick made with biogas as an energy source. Biogas is a natural energy originating from the anaerobic combustion of organic matter (food waste and forest biomass). Using this type of heat, the product has a lower embodied energy over other kinds of bricks. According to the company, CO_2 emissions are reduced by 35%, equivalent to 17,000 tons less CO_2.

The Ecoclinker model is available in two sizes: 10.6 x 5.2 x 1.8 in. (27 x 13 x 4.5 cm) and 9.4 x 4.5 x 1.9 in. (24 x 11.5 x 4.8 cm). Its resistance is 3.6 ton/in^2 (55 N/mm^2) and can be applied in ventilated and self-supporting façades and in external walls of the housing plot.

The EcoManual, also manufactured by Piera Ecocerámica, can be used for exposed interior walls e.g., in the living room. It is manufactured with significant energy savings and CO_2 emissions, as it uses biogas as a source of heat for baking the bricks.

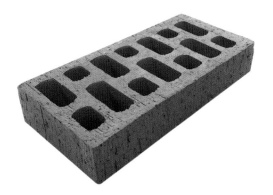

Thermoclay is a low-density ceramic block. Lightening agents that are gasified during the baking process at over 1650ºF (900ºC) without producing waste are added to a clay mixture, and an evenly distributed fine porosity forms in the body of the block. The construction of the product and studied geometry of the piece means that one-layer walls have equivalent or superior performance in some respects to multi-layer walls. This material offers good thermal and acoustic insulation and enough mechanical strength to be comfortably used as a load-bearing wall in parts of Europe.

The latest Termoarcilla (thermoclay) ECO technology is specially designed to give optimum thermal insulation to blocks that incorporate Thermoclay. The images show two different Termoarcilla ECO morphologies.

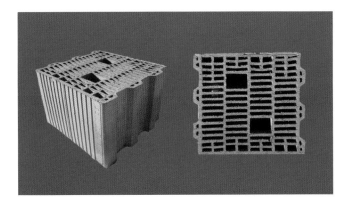

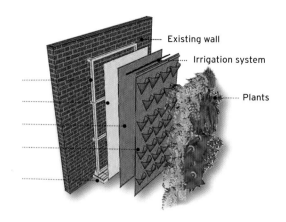

Existing wall

Irrigation system

Plants

Green wall

A green wall cannot be considered as a structural element of the house, but as an aesthetic and functional accessory, as it provides thermal and acoustic insulation, in addition to purifying the air in the house. It can also be used as a partition with both sides covered in vegetation.

These systems consist of a structural support, fed by a controlled irrigation system. The plants grow on a hydroponic medium that does not require soil and is designed so that moisture does not affect the wall on which it is supported. Depending on the species used, a green wall can take six to eight months to acquire its optimal appearance.

FLOORING AND WOODWORK

When choosing materials for the floor or woodwork of the house, the origin and environmental impact throughout their life cycle is important. When it comes to the woodwork, wise choices should be made. Woodwork that prevents the problem of thermal bridging should be selected, and also that supports the installation of an underfloor heating system fed by renewable energy. These choices can lead to significant energy savings and the optimum climate control of the house.

Wood is normally used for doors. In this case, and as in the windows, we have to make sure that wood is certified by the FSC or another agency, assuring the origin of the wood fiber.

Wood, aluminum and PVC are the three different materials that can be used in the joinery of windows. Of the three, only wood can be deemed satisfactory for an ecological home. The lifespan of wood in windows is 50 years; however they can be restored to last longer. Wood acts as good sound insulation and it maintains warmth in the home.
PVC panels, although cheaper, are not recommended as they contain chlorine. They can also react to the sun due to exposure to heat and ultraviolet light. Aluminum has a very high embodied energy (86,000 btu/lb. [200 MJ/kg]), while for wood it is only an average of 860 btu/lb. (2 MJ/kg).

Slate, besides being used as roof tiles, may also be used as paneling and flooring. If slate is used as flooring, it is recommended that it is used indoors, as it is robust and resistant to abrasion. If slate is used for a patio, it should depend on the location of the patio, as it does not withstand frost well.

For indoors, stone floors, marble, travertine and granite (mainly quartz) are recommended, although the high degree of radioactivity of these materials has been noted. Limestones, as they are chemically weak, should only be used for curbs and steps. Local stone should be used where possible.

Local wood should be used for floors where possible. The most recommended types of wood for floors and for furniture and joinery are fir, chestnut, cypress, beech, larch, pine, walnut, poplar, oak, maple, birch and ash

FSC certified wood, when it comes from sustainable plantations, acts like a solar energy accumulator. To maintain it in as natural way as possible, use low impact borax or varnishes.

Teak from Southeast Asia, Africa or Central America, is used for furniture and flooring. Although local wood consumption is always more desirable, if choosing teak, make sure that it has been legally imported. Greenpeace highlights in its reports that 30% of world imports are illegal.

When installing a bamboo floor, the boards must be left for two days in the space to adapt to the level of humidity. If the structure is overheated in winter, there is the risk that the material will become excessively dry due to the lack of ventilation in the home.

Bamboo panels can be used as floating parquet, solid floorboards and outdoor paving. The resistance of laminated bamboo for flooring is similar to that of red oak. Ensure that it has a certificate of sustainability.

Cork is a natural product that acts as a good thermal and acoustic insulation material. It is also resistant to parasites. It is extracted from the outer bark of the cork oak every 9-14 years; it is pressed and reduced into sheets suitable for covering walls and floors.

Cork flooring is warm, resilient and durable. Cork coated with artificial resin (polyurethane) or PVC finishes should not be used.

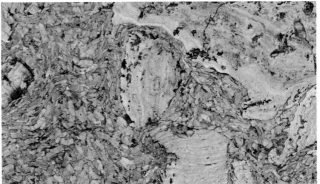

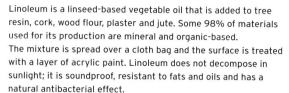

Linoleum is a linseed-based vegetable oil that is added to tree resin, cork, wood flour, plaster and jute. Some 98% of materials used for its production are mineral and organic-based.
The mixture is spread over a cloth bag and the surface is treated with a layer of acrylic paint. Linoleum does not decompose in sunlight; it is soundproof, resistant to fats and oils and has a natural antibacterial effect.

Ecoralia markets the recycled material Madertec, with a composition that is similar to that of wood. Hidden staples are used in its installation. This flooring, which is composed of recycled plastic from the municipal collection of packaging waste is especially suited to outdoor use because it is rotproof, does not chip or decompose, and is highly resistant to moisture.

The Ston-ker Ecologic by Porcelanosa is a porcelain tile made from recycled material. It is commonly used in large projects and civil works, but is also marketed for household use, as can be seen in this patio.
Some 95% of this ceramic coating is made from recycled material leftover from ceramic production processes, without losing the strength and versatility of porcelain tile.

PASSIVE AND ACTIVE STRATEGIES FOR THE ENERGY REQUIREMENTS OF AN ECOLOGICAL HOME

A photovoltaic solar energy installation is formed by solar panels with two or more semiconductive layers, usually silicon, which generate electrical charges when exposed to sunlight. Its production consumes a lot of energy, so it takes an average of five years to offset the carbon emitted during its manufacturing process. If the house in which the panels are installed is off-the-grid, the system includes an accumulator battery. Other circuit elements regulate the load and convert the 12V that the panels supply to 220V and 50 Hz alternating current for domestic consumption.

Typically, installation is regulated by each country's specific technical rules of construction and by the criteria for architectural integration of local ordinances.

The energy requirements of a home are measured in kWh. One kWh is equivalent to running a 100 watt bulb for 10 hours. Apart from the bioclimatic design, the market offers a variety of technologies, within the active strategies, that help meet the energy needs. In the event of surplus, and in line with the regulations established by each country, surplus energy can be sold to the grid at a price set by the concerned government. The operational diagram of some of these technologies has been described in Chapter I. Below we will further explain their technical characteristics and applications.

The main complaint regarding the installation of solar photovoltaic panels is the space required.

Its expected useful life is 25-30 years. Per kilowatt installed the installation costs $7,000 to $14,000. It is cheaper if all the energy produced is consumed, otherwise you need to install a meter that costs $500 to $600. Find out if there are grants available for their installation.

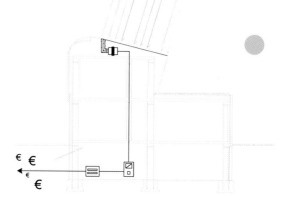

If no anomalies occur, a photovoltaic solar panel is cleaned with soap and water. According to where it is installed (gabled roof, façade or flat roof), special cleaning staff may be required.

Solar thermal energy has been used since the 70s and harnesses the sun's energy to produce hot water and heating. Such facilities consist of special panels, called collectors, which concentrate and accumulate solar heat and transmit it to a fluid that we want to heat. This fluid can be either the household drinking water or the hydraulic system of home heating.

Typically, its installation is regulated by each country's specific technical rules of construction and by the criteria for architectural integration of local ordinances.

The most widely used solar collectors at a household level are flat plate collectors. A flat solar panel consists of a box insulated on the bottom and sides and a thermal absorber plate, a metallic plate, which is attached to this insulation. The liquid to be heated flows through the pipes welded to this plate.

In recent years, vacuum solar collectors, which are more expensive to buy, have emerged. They better integrate into the building and are more efficient because they do not use pipes with antifreeze; an evacuated space is left in the gap between the protective glass and the absorbing surface. The evacuated tube collectors are cylindrical.

This model, sold by Chromagen, presents its Thermosiphon system for assembly on flat or sloping roofs. The system consists of a storage battery (79 gal.) with a double-envelope heat exchanger, and two flat collectors with a selective absorber. It is installed on a steel modular support protected against corrosion. The tank conserves hot water generated by the collectors for a limited time, between one and four days in the case of small systems.

Some installations with flat plate collectors have an engine room in the basement of the house, where the hot water storage tank and a high-efficiency boiler and auxiliary system are stored. Kyoto House by Pich-Aguilera Arquitectos also incorporates a dual-flow heat recovery system with the controlled renewal of air.

A solar thermal collection system has a lifespan of 20 years, after which it will require an upgrade. The initial investment to install it in a single family home is recovered in five years. Such facilities can feed a system of underfloor heating. If you have a heat pump, it can also be used for radiant cooling.

We often associate wind energy with wind turbines that we see in the countryside, but this type of renewable energy can also be used for small-scale residential buildings. There are two types: those that are placed on the roof and those located on masts. The former are capable of charging 12–24 V batteries from 100 W to more powerful models with 2.5 kW turbines that can export their surplus to the electrical grid.

Mast turbines are 10–50 ft. (3–15 m) tall with a capacity that can range from 600 W to 20 kW. Its actual energy production depends on the length of the blades of the turbine, the wind speed and whether the turbine is obstructed by a building or mass of trees.

If we assemble a mast-type installation, the best results are achieved when it is 32 ft. (10 m) or more away from the buildings and surrounding trees. It is important not to install a turbine at a point where there is excessive turbulence.

The majority of single-family homes have a consumption of less than 15 kW. To budget for an installation of this type, a rough calculation of $4,000 per kilowatt installed should be made, which will include the turbine, the mast, the inverters and the batteries (compulsory if the home is off-the-grid). If you choose to sell energy to the grid, you must install an export meter, which costs roughly $500 to $600. Find out if there are grants available for its installation.

An annual review must be carried out to check for any damage. Microturbines last 15 to 20 years, so the depreciation during their lifetime always depends on the purchase price that the government establishes for surplus sold to the grid.

The Energy Ball V100, sold by Home Energy International, has a power output of 500 kWh, and an average speed of 23 ft./s (7 m/s). It can be assembled on roofs with a mini-mast 15 ft. (4.5 m) high or with an independent mast measuring 32–39 ft. (10–12 m).

Geothermal energy has been used since the mid-70s in countries like Switzerland or Finland. It consists in the use of the Earth's internal heat to supply hot water and heating to buildings or infrastructures. In the case of residential housing, these types of installations use the gradient temperature of the surface layers of the earth, as from a certain depth the soil temperature is constant (60°F [15°C] in Mediterranean latitudes). Therefore it can be used to cool down structures in summer, and as a form of heat in winter using a heat exchanger.

The external circuit, called the collector, is a plastic pipe circuit through which an antifreeze solution flows. It should be installed in the design phase, during the earthmoving stage, because otherwise the price increases. For a 10 kW installation in a 1,300 ft^2 (120 m^2)house the cost is approximately $18,000. For every 10 ft^2 (0.9 m^2) of housing, 3.5–6 ft. (1–1.8 m) of land is required. Trees cannot be planted on this land. Therefore, horizontal geothermal installation above all requires space.

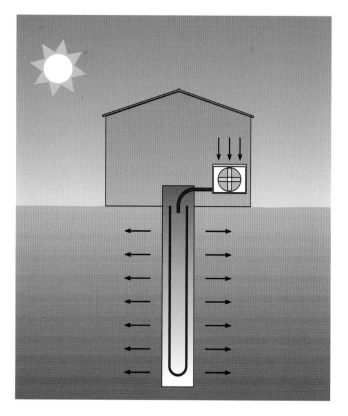

Vertical installation is more expensive because of the need to drill up to 330 ft. (100 m) deep. It is ideal for urban environments. Inside the home, using either system, climate control can be in the form of underfloor heating, air convectors or conventional radiators. In a humid climate, underfloor heating is recommended, because on a cool floor, a film of water may form due to condensation.

A geothermal system can generate savings of up to 80% compared with oil and 70% with gas.

Heat-energy combined cycles are dual systems that produce electricity and heat and reduce energy consumption by 25%. As they are domestic systems, electricity is consumed where it is produced, thus there is little loss due to transfer. These combined cycle devices are designed to operate during the day and release a consistent low heat, and the cost is recouped in three to four years. Their price ranges from $650 to $1,300. A standard model can produce approximately 2,400 kWh of electricity per year and approximately 18,000 kWh of heat.

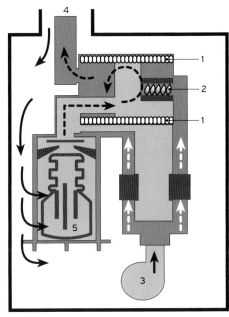

Passive strategies derived from the bioclimatic design of a home are those that consume less energy and are most beneficial to the environment. Mainly, they consist of achieving a high degree of thermal insulation, they take advantage of the orientation of the building and the greenhouse effect as a source of heat in winter, and they facilitate cross-ventilation to cool the space in summer.

1. Heat exchangers
2. Auxiliary burner
3. Combustion fan
4. Balanced flue
5. Stirling engine and alternator

In Europe, *Passivhaus* certification exists, a seal originally from Germany and Austria, which summarizes the main requirements for a bioclimatic profile house:

- high level of isolation and reduction of thermal bridges
- high levels of solar gain
- good air quality inside the home, using a mechanical ventilation system with heat recovery
- excellent level of airtightness

If the aforementioned points are complied with, the housing virtually requires no additional system of heating or cooling generated by active strategies which, as we have seen, consume higher levels of energy. If you require any auxiliary systems, install an integrated system for heat, hot water and ventilation.

During winter in temperate climates or throughout the year in cold climates it is useful to take advantage of the greenhouse effect. The design of the house should include galleries in the south façade. In this context, higher thermal mass flooring such as concrete facilitates the accumulation of heat, to then be released inside when the temperature lowers.

ROOM CLIMATE CONTROL

One of the clear advantages of the bioclimatic design of a home is that it can control the interior temperature naturally. When this alone is not sufficient, mechanical systems can be used to provide heat or cool air. This section presents solutions for zero or low-energy consumption climate control.

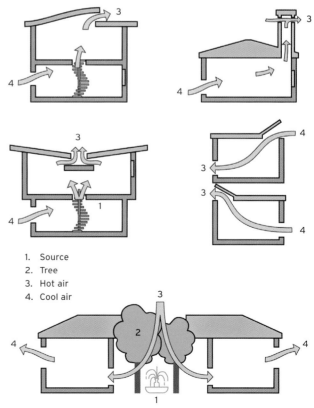

1. Source
2. Tree
3. Hot air
4. Cool air

The interior courtyard with trees and vegetation serves as a source of fresh air to cool rooms in the house in warm climates.

In this case, instead of a courtyard, there is a gallery that connects the two wings of the same building. As it is glass and due to the lush vegetation inside, it generates cool air thanks to the evapotranspiration of plants and the effect of cross-ventilation that occurs between the two sides of the house.

A rooftop garden is another winning choice to naturally control the temperature in a house. The vegetated cover acts as a thermal mass that regulates the temperature inside. It has a dual effect: in summer it prevents the roof from overheating, while in winter, it counteracts the cold as it accumulates heat during the day. It also provides a home for birds and insects and blocks out noise, by reducing the noise inside up to 8 dB.

A nearby wooded area with deciduous trees also has beneficial effects, especially in hot countries. In the summer it blocks out the sun, while in the winter it lets the sun through.

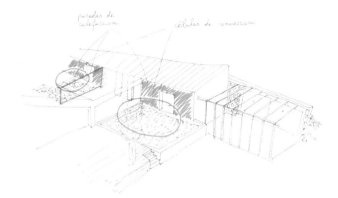

The generation of cool air by evaporation is also beneficial, although not recommended for humid climates. This house in Spain is surrounded by two pools that cool the nearby rooms through evaporation. In winter the water accumulates heat, which then passes to the walls of the high thermal inertia house.

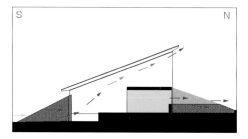

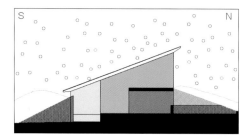

Cross-ventilation systems are recommended for moderate climates. They are more efficient in the case of isolated dwellings since their effect is reduced if you live in a compact city.
It consists of fresh air circulating through the rooms. Lantern windows, skylights and other openings that connect the coldest part of the house with the hottest are excellent to enhance this current.

This Shimane house (Japan) leverages the fact that it is half buried between stone gabions so that air can circulate between the crushed stone and the gaps that draws the roof structure.

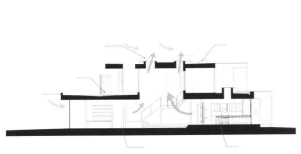

In this house in Santa Monica, California, the solar chimney effect allows air to exit through the stairwell by two motorized skylights in the roof, which expel the hot air in summer and helping fresh air to enter through the ground floor

Concrete is not an eco-friendly material. However, the use of concrete along with heat accumulators such as a Trombe wall can act as a natural source of heat inside the house. Concrete is one of the materials with the highest thermal inertia. The heat that it collects from sunlight during the day is released at night as natural and free heat. Obviously, the more the house faces south, the more heat it will accumulate.

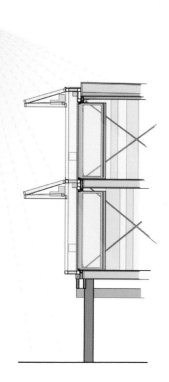

In hot climates, large eaves protect the house from sunlight, to maintain a cool interior. Shutters with air chambers between them and an interior wall also carry out this same function, as do awnings, and the elaborate system of skins used in this house in Chile.

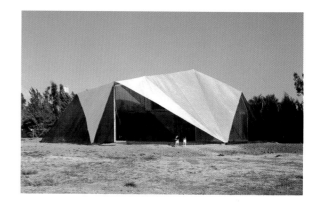

A bio-house must comply with many requirements including window insulation, load-bearing walls and roofs, and thermal break.

This diagram shows the heat loss from a house without insulation:

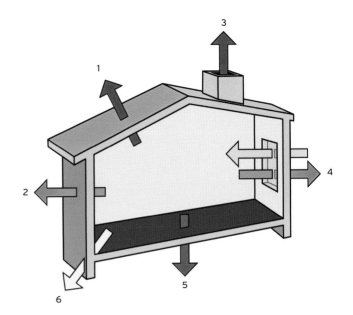

1. 25-30% (ceiling)
2. 20-25% (walls)
3. 20-25% (renewal of air)
4. 10-15% (windows)
5. 7-10% (floors)
6. 5-10% (thermal bridges)

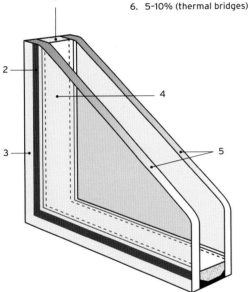

1. 39% of solar gain
2. 70% light through the glass

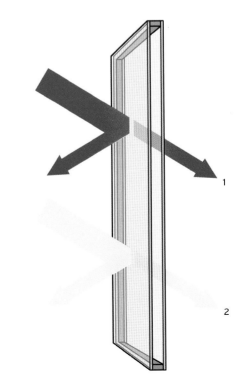

To prevent this heat loss, the windows should have double or triple glazing, which prevents unwanted heat accumulation or loss as well as blocking out exterior noise. The sketch below shows standard double glazing:

1. Molecular sieve absorber
2. Primary sealant (vapor barrier)
3. Secondary sealant
4. Profile separator
5. Glass according to requirements of strength, security and transmission properties

Moreover, the quality of the glass also influences the amount of sunlight that enters the house. Low emission glass (Low-E) is ideal for warm climates. It reduces solar gain and allows visible light to pass through the glass into the building. In the example, the reduction percentage is given for this particular case.

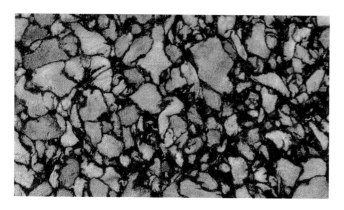

When choosing insulation for curtain walls, opt for natural and biodegradable insulation that regulates both thermal comfort and water comfort. The most recommended forms of insulation are cork, hemp, coconut fiber, clay, straw, lime, expanded recycled paper, natural wool, shavings and expanded clay, perlite and vermiculite (feldspars and expanded rocks) if they are well sealed. Value higher than K, less isolation. Below, we explain the most important types of natural insulation.

Cork chips are used to fill cavities, as pressed cork panels, to be used in roofs or to cover surfaces. They are extracted from the bark of the cork oak. K value: 0.045 W/(m°K).

Hemp is a natural fiber that grows rapidly and is easily cultivated. It is used to make natural and breathable insulation blankets. K value: 0.041 W/(m°K).

The insulation properties of wool improve when wet. If you have a local supplier and manufacturer, it is a good option. K value: 0.04 W/(m°K).

Wood fiber panels usually are made from waste from processing wood or small branches, therefore its use is compatible with sustainable forest management (FSC or equivalent). Panels are sold with thick resinous wood fibers chipboard with plaster or white cement or light panels with tiny fibers. Both have the same K value: 0.05 W/(m°K). The light, porous wood fiber boards can be used in partitions, façades, roofs and floors.

Cellulose insulation is made up of waste paper and is used to insulate air chambers. Although it needs to be treated with chemicals to prevent mold and protect it from fire, it has excellent insulation properties, it is lightweight and requires little energy to manufacture. Used loose, its K value is 0.0071 W/ft.°F (0.042 W/m°K), or, injected with a hose, its K value is 0.0066 W/ft.°F (0.039 W/m°K).

Finally, it should be noted that straw, a material known for its insulating capacity and thermal comfort, can be used in the form of straw panels and plaster for indoors.

Synthetic insulation such as rock wool, glass wool, extruded polystyrene and polyurethane, should not be used as they require more energy to be produced and are pollutants.

The ideal temperature for rooms is between 61°F and 64°F (16°C and 18°C). For every degree over 64°F between 250 and 500 lbs of CO_2 is emitted.

Following passive systems, *underfloor heating* is one of the more environmentally friendly ways to heat or cool a house. This feature consists of a grid of cross-linked polyethylene pipes that distribute the heat or cold and a distribution header. It is a type of "invisible" heating that generates less airborne dust and provides a stable temperature. Its critics suggest that if used in summer, it can give the feeling of cold feet and hot head.

Electric underfloor heating is not advisable, as it breaks down easier and consumes more energy.

A radiant heating system can also be used in the ceiling or walls in the form of panels. It is less common, but there are companies such as Runtal that sells models like the one in the image. Both systems are fed with hot water and can be heated by a solar thermal installation.

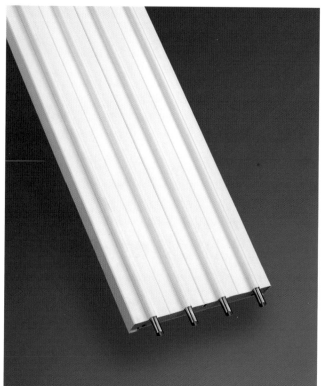

Another ecological way to heat your home – without using non-renewable sources such as gas or oil, or renewable sources such as firewood – is to use a *pellet stove*. The first domestic stoves of this type appeared in the U.S. and also have been used in residential complexes, proving successful in Sweden, Germany, Austria and France.

Pellets are a 100% biodegradable material obtained from recycling chips, splinters or sawdust from the agro industry. These waste products undergo a process of crushing, drying and pressing which results in flammable sticks. Pellets have a calorific value of approximately 2,000 kcal per pound (4,500 kcal per kg).

Chimneys have a vent measuring 3 ⅛ in. (8 cm) diameter. The fire is controlled by an electronic thermostat. The ashes generated during combustion can be used as a fertilizer for plants.
The accumulation of pellets at home is not dangerous, nor do they produce unpleasant odors. To produce about 1,800 kWh, it will take about 800 lb. (360 kg) of pellets. If using another type of biomass such as wood chips, the calorific efficiency is lower.

The installation price is $6,500 to $12,000. The economic advantage relative to an oil stove, which is much more polluting, is that the price of pellets is very stable.

Its maintenance is easy: periodically removing ash and cleaning the burner once a year.

Home Energy International sells a pellet stove that can be installed in the boiler room, garage or basement. It occupies approximately 160 ft² (15 m²), but requires additional space for the renewal of air and to store pellets.
Pellets can be fed mechanically or manually. As a heat source it can be used for underfloor heating, heating water and rooms via conventional radiators.

Just like underfloor heating, water radiators also use water as heating fluid. They have been used for many years; therefore companies have tried to improve their efficiency in recent years. One example are the Jaga Low-H_2O radiators, by the company Jaga, which consume 12% less energy. Such devices can be powered by a heat pump, a solar thermal installation or by condensing boilers. Some models come equipped with a wood casing, as shown in the photo at right.

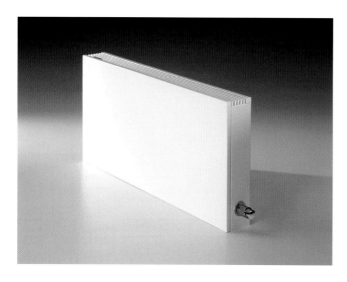

Jaga models may be accompanied by a manual or automatic ventilation system. This ventilation system is integrated into the Low-H_2O consumption radiator and maintains cool rooms at night. This ventilation device is powered by electricity.

These boilers burn fuel to heat a fluid (water) which then distributes this heat. If the fuel is gas, a gas boiler produces about twice as much CO_2 than electricity.

In the case of choosing a gas boiler, select a condensing boiler, which can represent a 40% saving in CO_2 emissions over the conventional boilers; these use heat and gas combustion itself, so they reach an efficiency of 98%. The most efficient systems are those that have individual temperature control in each room, or where the temperature for different times of day can be programmed. The use of thermostatic valves in each hot spot will control the temperature automatically.

The Saunier Duval Helioset boiler supplies hot water to the circuit of the house. The models consist of a collector and a reservoir (40 or 66 gal.). They both incorporate an automatic drain system, which prevents frost and the overheating of the solar fluid.

The performance of a geothermal heat pump can be 50% better than in the air-air heat pump cooling systems. With damp clay soils, the heat transfer is better than with sandy or dry soils. These pumps feed the underfloor heating (heating and cooling) and hot water system (DHW). In heating, savings can amount to 75% and cooling, more than 80%.

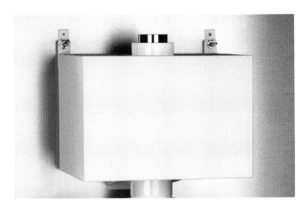

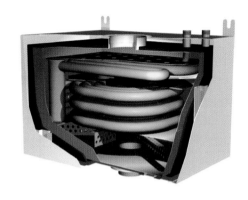

The most efficient gas boilers are "combi boilers," which heat water on demand so you do not have to heat the entire tank of water. It is also advisable to properly insulate pipes where the water is distributed to reduce heat loss.

The image shows the Zenex device that is placed on the top part of the combi boiler, which extracts residual energy from the gases and preheats the cold water supply. The price of this device is about $750 plus installation costs. Up to 1,300 gallons of water can be saved, while gas consumption is reduced by 11%. The reduction of CO_2 emissions from the house amounts to approximately one ton per year.

If cross-ventilation is not enough, fans are always a better option than air conditioning. These smaller devices consume less energy and do not require chemical coolants.

The Turbo Air 3200, manufactured in Canada, is two to three times more efficient than other conventional fans. According to the producer, its consumption is less than a 100 watt incandescent bulb. It has three speeds and is noiseless. The revolutionary design that enhances the Venturi effect distributes air and envelops the user, so to speak. It can also be wall mounted.

Heat pumps use energy to extract heat from outside air and transfer it inside. In terms of efficiency, it is one of the best systems, for each kWh consumed 2.5 to 3 kW in hot or cold air is released. The machines with inverter technology are especially efficient, as they have an electronic control, reducing power consumption by more than 30%. Make sure you choose a machine with energy efficiency class A and your home is well insulated. However, heat pumps are not suitable for extremely cold climates, as they are less effective.

Home automation control systems adjust the temperature of a home according to the variation in the outside temperature, the time of day, the area of the house or the presence of people. The intelligent automatic control of awnings, blinds and curtains allows or prevents passive solar gain according to the thermal requirements of each season. The system also alerts us if it detects any inefficiency in the opening or closing of windows.

HOW TO SAVE WATER

Drinking water is only guaranteed for 67% of the world population and some 20% do not even have access to this resource. However, the use of rainwater or graywater purification (washing machine, kitchen and shower) is not covered in any construction regulations, even in countries where this resource is scarce.

A system for collecting rainwater should be planned in the design phase, as it requires the installation of a separate collection network for the bath, shower and sink drainpipes. The installation price is $2,500 to $4,000, and the payback period, depending on the price of water in each country is 10 to 15 years. Water savings can amount to 30-45%.

The diagram represents the rainwater collection system, the collection room for this water and the distribution system of the graywater for Kyoto House by Pich-Aguilera.

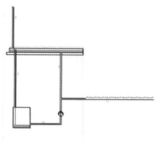

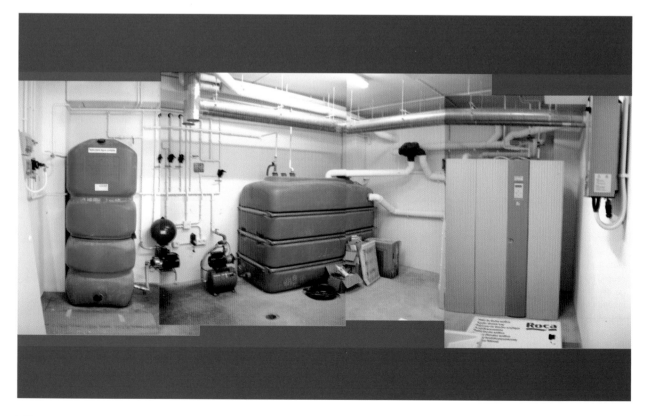

The image shows a tank, which collects rainwater that can be used for the toilet, washing machine and to water the garden. If the house has a roof garden, this garden stores between 50% and 90% of the rainwater that will gradually yield to evapotranspiration, while the remainder could be collected to be reused.

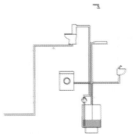

The diagram shows the graywater separation network for Kyoto House, from the washing machine, shower and sink.

Hansgrohe sells Pontos AquaCycle systems, which processes graywater from the shower and bath through a mechanical and biological process in which clean water is produced that can be reused for the toilet, laundry, household cleaning or watering the garden. This system will halve the consumption of drinking water but it needs to be separated from the sewage network, which is the toilet water that is sent directly to the sewer.

The simplest (AquaCycle 900) system is composed of three 80 gal. (300 l) tanks. After a pre-filtration process, water is subjected to a double biological treatment and then to a UV disinfection. Energy consumption is 1 or 2 kW per cubic meter. The result is recycled clear water that meets the EU directive on bathing water.

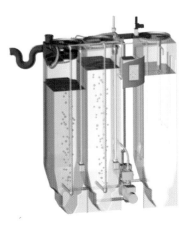

LIGHTING SYSTEMS

Before deciding on the most respectful lighting system, we should ask ourselves whether or not everything possible has been done to maximize the natural light that reaches the site in the design phase of house. We recommend that rooms where you spend most of the day face south, and if you need nonstructural partitions, use glass block walls that let light through.

For points of the house facing north or blocked from the sunlight by trees or large shrubs, there are other solutions on the market such as solar tubes. This type of tube helps to redirect the light where we want it. It is usually attached to the roof through a porthole, and through an articulated and reflective tube light is directed to the ceiling of the room where it is required.

La Bonne Maison (France) is an example of a house with optimum solar orientation: some 72% of natural light enters through the south façade.

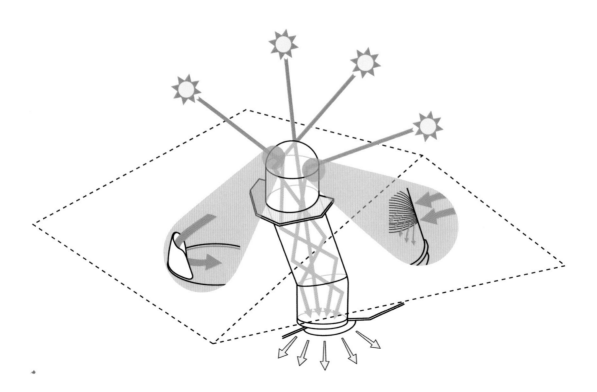

Both the design that enhances the orientation and the use of solar tubes for shaded rooms exemplify the constructive measures of low CO_2 emissions.

Skylights are also a useful resource to take advantage of maximum solar radiation in some areas. In this home, the inhabitants enjoy a natural skylight that floods the kitchen with light.

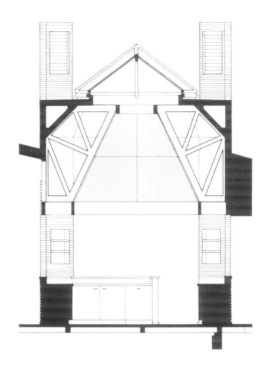

At other times, such as in this prefabricated home in California, the architect proposes lantern windows to help gain natural light in the north façade, which in this case is where the kitchen and storage lockers are located.

When the electrical installation is being carried out, recessed halogen bulbs should not be used. While they offer the best quality of light, they are contaminants due to their high power consumption. Another prime example of inefficiency is the incandescent light bulb.

There are two options, therefore, that should help when selecting the type of electrical installation to be installed: a) low-energy lights that can be used in the same wall lights where incandescent bulbs were used, and b) LED modules and fixtures.
LED, or light-emitting diodes, were first sold for industrial use. Currently, their domestic use has gained favor, thanks to technical innovations and because they are more efficient than light bulbs.

Parathom Classic by Osram are LED bulbs designed to replace incandescent bulbs up to 40 W. They have a lifespan of 25 years and have energy savings of 80%.

Compact fluorescent bulbs are available without the shell with the exposed twisted conductor or in the shape of a bulb. They are sold with screw cap and various power and light tones. Consuming 80% less energy and generating 80% less CO_2 they have a 15 times longer life.

This model, Dulux Carré by Osram, can be used indoors or outdoors, as it is waterproof. The model can be equipped with a sensor that, according to the amount of light outside, decreases or increases the light output. Three models are available: 9 W, 11 W and two 11 W bulbs.

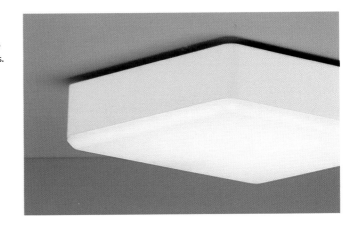

Besides using low-energy lighting, it is also important that the system itself is efficient. Home automation systems adapt the lighting level depending on the variation of sunlight, the area of the house, the presence of people or the time. Intelligent automatic control is also available for the home's awnings, blinds and curtains, so to make better use of sunlight according to the time of year, which will also have an effect on the consumption of the home's energy.

HOME AUTOMATION LIGHTING CONTROL DIAGRAM

A. Sensors
1. Light detector
2. Presence detector
3. Time of day

B. Controller
4. Home automation control

C. Actuators
5. Voltage regulator actuator (lighting)
6. Voltage regulator actuator (on/off)

D. Interfaces
7. Keypad
8. Mobile / SMS
9. Switch
10. Internet

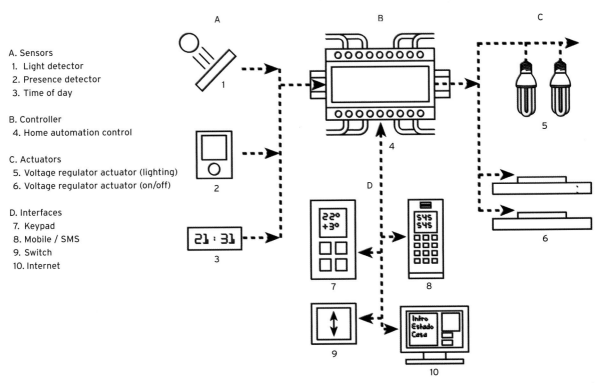

HOW DO I CONVERT MY HOUSE INTO AN ECO HOME?

Better late than never. Most actions outlined in this chapter are permanent solutions that will help save energy, water and resources and reduce waste, and that can be implemented even in homes that were built many years ago. These measures are mainly related to surfaces, materials and furniture inside the house. Let's take a room-by-room look at the house, from the living room to the bathroom, including the yard trying to preserve materials so that they have a longer life. The longer our materials and articles last, the better for the environment.

The use of glass expanses guarantees the entry of light and passive solar gain depending on the orientation of the living room, as well as energy savings from a reduction in winter heating and artificial lighting consumed.

LIVING ROOM AND DINING ROOM

The dining room and living room are daytime spaces that require good light and ventilation. If wood is chosen for the furniture and floors, it should be locally sourced. Materials such as mahogany, redwood, teak, rosewood and ebony should not be used as they grow far from where they are consumed and their sustainable management is not always guaranteed.

The kitchen and dining room often open onto the living room. In this case, apart from benefiting from the natural lighting from the glass expanse, it also takes advantage of the high thermal inertia of the flooring (concrete), which accumulates heat during the day and releases it during the evening and night.

The kitchen and dining area share the same space on the first floor of this prefabricated home in Belgium, inspired by greenhouses. The façades are made of highly insulated glass panes and translucent polycarbonate panels, which are also insulated according to the front façade.

The dining table is composed of a slab of Belgian blue stone, a jarrah frame – an Australian variety of eucalyptus – and French oak legs. The wood from the frame has been recycled from the floorboards of an old ship and the legs had been previously used in the construction of a former brewery.

The floor is pigmented concrete, of which the thermal inertia can be used. Below the concrete flooring there is a 4-in. (10 cm) layer for the installation of the heating/cooling ducts of a horizontal geothermal heat collector.

One of the façades of this environmental education center located in Bainbridge Island, Washington, features a projected roof to reduce the amount of sunlight that enters. In contrast, the adjustable windows in two of the façades create cross ventilation during hot days.

The use of bamboo laminate flooring in this house in Novato, California, is an alternative to traditional wood flooring. When buying bamboo, ensure that it is at least six years old. If it is cut too young, it is not as hard.

A gap left below this type of floorboard means that it allows for the installation of electricity, air, water (underfloor heating/cooling) and voice/data lines.

For good maintenance, humidity should be consistent across the boards. Otherwise, the flooring may suffer deformations. If synthetic varnish is used, it should not contain aluminum oxide, or it will leave very visible white marks if scratched.

The flooring and the wooden table in this property are treated with natural oils and waxes manufactured by the company Auro. They do not emit volatile organic compounds (VOCs) and ensure the long-term maintenance of the properties of the wood.

If you lay rugs in any part of the house, ensure that they do not contain brominated flame retardants. These bromine compounds that retard the ignition of the product can be found in the plastic used in television sets and other electronic applications.
Rugs that release volatile organic compounds (VOCs) should not be used.

Gabarró sells this FSC certified solid natural wood floorboard. Here, jatoba wood has been used, but they are also available in cumaru, ipe, sucupira and teak. The problem with these materials is that the wood has not been locally sourced.

Acrylic plastic varnish is used for the finish, which is not the best environmental option on the market. The UV finish repels most domestic liquids and facilitates its long-term maintenance when subjected to direct sunlight.

The Karelia single-plank flooring, soldby Gabarró, with Rock Salt oak in this case, is PEFC certified under the category 3 group C (certified chain of custody). This type of certification is regulated by wood producers; therefore it is not as restrictive as the FSC certification.

Listone Giordano sells hardwood floorboards with natural oil-based Natif varnish finishes. These oils are applied once or twice a year, depending on the level of traffic and areas worn from use. Wood floors adopt a look of old leather and an effect of the oil finish, and are certified by the FSC and PEFC as appropriate. The wood comes from species like oak (Fontaines), the cabreuva (South America) and teak (Asia).

Wood floors support underfloor heating. In this case, the system consists of planks of plywood with underfloor heating (hot/cold).

The solar house III was designed in Switzerland according to the criteria of the certification of *zero energy* housing. This type of construction is not connected to the grid and has a net energy consumption of zero; i.e., it generates the energy it consumes. In this image of the living room, the prefabricated wood construction can be seen as well as the energy accumulator panels on the façade, formed by a type of paraffin that is heated or cooled according to the outside temperature.

This table, made by Fusteria Ollé, is manufactured with bamboo *Density Mombasa* and finished with a deep matte water-based varnish.

Fusteria Ollé manufactures bamboo windows inside and aluminum or bronze on the outside. The interior and exterior cladding gives these windows the appearance that they are hermetic and high security.

Deutsche Steinzug markets the ceramic floor tiles in the image, which are mainly made of clay, kaolin, feldspar or quartz. They are suitable for allergy sufferers. All materials used are locally sourced and are produced near the quarry where they are extracted. This German company guarantees a clean production process, in which the remains are reused, waste water is purified and waste hot gases of the process are reused to preheat.

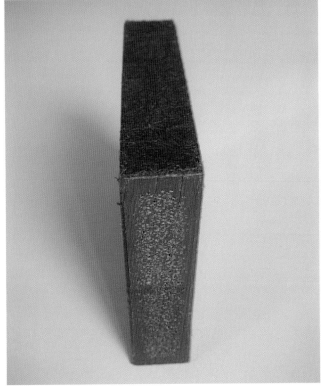

The Barcelona-based company Zicla markets these recycled plastic baseboards. The plastic, mainly polystyrene and polypropylene, comes from selective collection containers.

KITCHEN

The kitchen is another daytime area where health and sanitation should be taken into account. We should ensure that the furniture does not contain formaldehyde, especially if there is conglomerate board, particle board, plywood or MDF (medium density fiberboard) in your furniture (cupboards, shelves and storage areas). These items contain a binder urea-formaldehyde resin, which binds the fibers with wood veneers.

Another important aspect is the use of modular furniture because of the space that it saves and its multifunctional character, which in the long run represents a reduced exploitation of natural resources.

If we aim to have a formaldehyde-free kitchen, the best advice is to use furniture and surfaces with natural or reused materials. In this example the kitchen cabinets are handmade using Zebrano (or Microberlinia wood imported from Central Africa), and with French vine stalk handles.

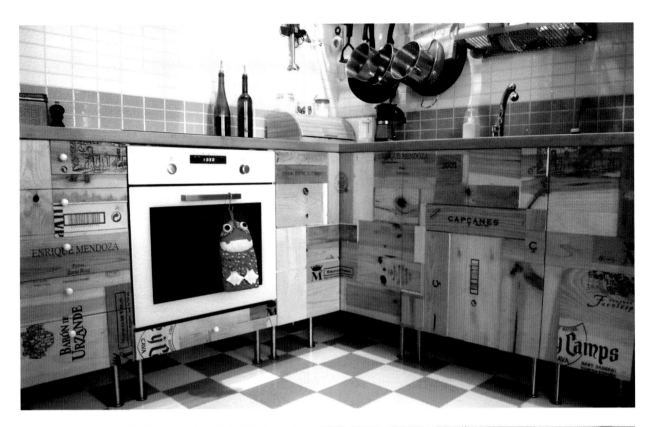

The R3project, conducted by the eco-designer Petz Scholtus and with the help of the industrial engineer Sergio Carratala, consists of the ecological reformation of an old apartment in the Gothic Quarter of Barcelona. In the kitchen, for example, the paneling of the furniture is made of wood from used wine boxes. The worktop is untreated FSC certified solid wood. A natural oil finish has been applied for the correct maintenance of the worktop.

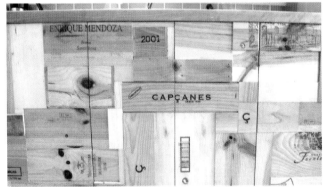

The Fujy project in El Escorial (Spain) presented the latest technologies in sustainability. The modern lines of this kitchen include a worktop with recycled materials and class A appliances. Electric burners should be powered by renewable energy.

This prefabricated home has a kitchen with FSC certified wood and formaldehyde-free kitchen furniture. The worktops contain recycled paper and all appliances are Class A or higher. The interior floors are covered with laminated bamboo floorboards.

When choosing a worktop, choose granite or its artificial imitation, Silestone. Granite is heat resistant, but a lot of waste is produced from the quarries. Silestone has a good aesthetic quality, but it is composed of quartz, resins derived from oil and the antibacterial agent Microban (triclosan), therefore it can be deemed an antibacterial worktop but not entirely eco-friendly.

Quartz is a very hard yet porous mineral rock, so you should be careful not to stain it. It is composed of 90% calcium carbonate and contact with acidic agents should be avoided.

If you choose a wooden worktop, make sure that the fibers come from a sustainable plantation. Beech or solid oak are recommended, and the most reliable certificates are from the Forest Stewardship Council (FSC), the most rigorous accreditation organization in this sense. Solid wood is always advisable over plywood or conglomerate boards.
To enhance its durability, good maintenance is required. We recommend you to varnish it with natural oils to help protect it from moisture.

Stainless steel worktops are clean and safe, and for hygiene reasons they are the best choice for professional kitchens. In contrast, they have a cold appearance and a lot of energy is required to produce them.

Ceramics cannot be regarded as 100% ecological, as their development process requires a lot of energy. They are made from clay, feldspar and sea salt at temperatures of 2400ºF (1300ºC). They are used both in kitchens and for floors.

If you use bamboo as a solid wooden board, make sure that it comes from forests with controlled felling and replanting processes.

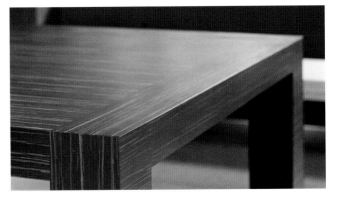

Grupo Cosentino manufactures the worktop ECO by Cosentino, which can be used for kitchen countertops, bathrooms, walls and floor tiles. It has very low porosity and high durability; it does not require sealing and is resistant to stains, scratches and heat. This product is certified by Cradle to Cradle (C2C) and Greenguard. The Greenguard Certification certifies that the product has no adverse impact on the air quality of the home where the product is used. The C2C seal ensures that when the product comes to the end of its life cycle, it will be reintroduced into the industrial cycle without generating waste material.

ECO by Cosentino is made of 75% post-industrial and post-consumer recycled materials and 25% natural elements. Among the former are glass, mirrors, porcelain and vitrified ash. Among the latter are traces of quartz and a common resin composed of 22% corn oil.

The flooring and cupboards of this kitchen are made of hardwood. The south-facing walls store heat depending on the outside temperature. A prefabricated construction method has been used in this home. This is a system that generates less waste and reduces assembly time.

This kitchen features a central island in cherry wood. The floor and worktops are made of OSB chipboard. These boards are made of wood chips glued together with polyurethane resins and MUPF (melamine-urea-phenol-formaldehyde). These panels are an economical alternative to solid wood; however they contain chemical adhesives applied under high pressure and temperature, which release formaldehydes into the atmosphere.

The Evolution House kitchen, featured in the Alpexpo de Grenoble exhibition in 2008, features an American-style kitchen with OSB floorboards and a chestnut wood ceiling. Make sure that the OSB has low formaldehyde content and it has a certificate of sustainability for both materials.

This home located in Finland features spruce wood treated with natural oiled floors and walls. Cherry wood covered with a stainless steel worktop has been used for the kitchen furniture. The dining table is maple.

Both the flooring and the worktop are made of wood from the multinational company Listone Giordano. The wood is treated with natural oils and does not emit volatile organic compounds (VOCs).

The wooden floors and furniture of this kitchen are maintained using natural oils by the German brand Auro.

This kitchen designed by Delta Cocinas features solid bamboo from plantations with a controlled felling and replanting procedure, and a stainless steel worktop with flush LED. Stainless steel is a clean and safe material, but it has a high embodied energy.

The ceramic cladding for the floors, walls and kitchen of this house is made by Deutsche Steinzug, a German company that produces this highly energetic material in line with clean production criteria.

The use of natural light in the kitchen is fundamental to reduce energy consumption. Obie G. Bowman designed the central skylight to take advantage of sunlight. Sunlight filters through perforated metal panels (forming a truncated pyramid) that diffuse the light into the room, and some of it is reflected in the rooms next to the kitchen. The skylight also serves to extract hot air in summer and renew the interior air.

If the kitchen does not have access to outside, an efficient lighting option should be used. For many years fluorescent low-consumption and low-quality lights have been used in kitchens. Dulux Carré by Osram is an aesthetic alternative to fluorescent lighting using low-power indoor and outdoor lighting.

A green wall in the stairwell of this apartment in Berlin ensures a high level of indoor air quality due to the purifying effect of the plant biomass.

For cooking, the most environmental option is to use gas, flame or an electric burner. It is worth highlighting the use of electric burners or plates, unless the electrical system is powered by a renewable energy source. The more sustainable option of all is to cook by sunlight. The characteristics of a solar cooker are detailed in the section on accessories.

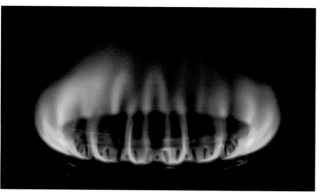

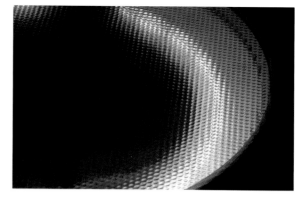

Natural Light is a water filter system for worktops that connects to the kitchen faucet and improves the taste, odor and color of the water for drinking and cooking. The key is a granular activated carbon filter and a high purity mineral patented formula.

This device removes 99% of chlorine as well as possible organic pollutants (pesticides, oils, dioxins), sediments and heavy metals. Indirectly, by avoiding buying bottled water, it reduces the amount of plastic waste generated.

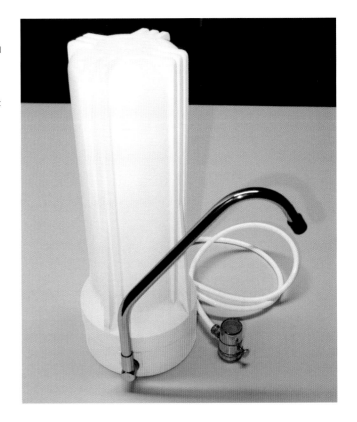

The installation of flow reducers on taps in the kitchen is a simple and inexpensive addition, and it reduces the flow by 50%. If the manufacturer provides environmental guidelines used in their production, all the better.

Implementing a home automation system can result in significant energy savings. For the kitchen, which includes many household appliances, appliances in standby mode can be detected and managed. This can represent between 8% and 10% of the total energy consumed in a home.

BEDROOM

The bedroom area is part of the private area of the home. The theories of feng shui converge in this space that ensures restful sleep, with our personal taste. This space should be neutral, it should be ventilated regularly, with the largest possible number of natural materials (furniture, carpeting, mattresses and fabrics), painted in light colors with water-based paints with little decoration on the wall. The presence of electrical and electronic equipment should be kept to a minimum.

This prefabricated prototype in Arizona, built by students under the direction of Jennifer Siegal, is used as guest accommodation. It has been laid out to facilitate ventilation along its rectangular construction. A central atrium separates the public area from the private area.

Waking up to natural sunlight in the bedroom, if curtains or translucent panels are not used, is good for our health. It also encourages cross-ventilation during the summer and periodically ventilates the room in winter, preventing moisture problems.

If using wooden flooring, ensure that it has an ecological sustainable certificate (FSC), that it is finished with natural oils, and is compatible with underfloor heating, as in the case of this Listone Giordano surface.

This house, built under the criteria of zero energy building (ZEB), has been 100% prefabricated, which can be seen in the wood panels that make up the interior walls, and the structure of the façade. Zero energy buildings have a net energy consumption close to zero in a typical year. They are self-sufficient renewable energy sources.

The carpet or rug used in the bedroom, as in all other rooms, should not contain volatile organic compounds (VOCs). The surfaces of the furniture should be smooth, and the wood should be certified by an international body and locally sourced. Beech wood is one of the most recommended types of wood.

Bright colors are not suited to the bedroom. To achieve maximum tranquility in this place of rest neutral tones should be used: white and a range of colors from ocher to beige. For total harmony, the quilt and sheets should be white.

Modular furniture in the bedroom is also advisable, for example beds that double up as a sofa, or in smaller details such as the television box that can be used both in the living room and in the bedroom.

Modular spaces are specifically tailored for small homes or student residences. As with the apartments versus family homes, reduced spaces are always more efficient and therefore have a smaller carbon footprint.

A foldout bed is an example of this, as it helps us make better use of living space. So, by day a table or something can be placed in the space or it can be left free, facilitating movement.

BATHROOM

The bathroom is where the most water is consumed. For example, whenever we flush the toilet several gallons of clean water are flushed down the drain. By now, the average citizen is already aware of the difference in consumption between showering and taking a bath.

Ideally, the water from the shower and the sink would come from filtered rainwater and water from the tank is purified graywater from the home.

The quality of water is also an aspect to consider: make sure water contains low levels of chlorine and no heavy metals, pesticides and volatile organic compounds, as ironically the drinking water network in large cities usually contains more than just water.

In the bathroom, underfloor heating or a radiator with water heated by a solar thermal system or a pellet stove is ideal.

The bathroom of this home on the Chilean coast stands out both for the natural character of the elements (sink, mirror frame and walls clad with wood and slate) and by the simplicity of its form. The lighting, especially natural, is complemented by a low-energy fluorescent lighting.

This house in Virginia has LEED Gold sustainable construction certification. The sanitary water is supplied by a geothermal heating/cooling system. Unfortunately, halogen bulbs are used in the bathroom lighting system, but in the majority of the home LEDs have been used, which are much more efficient.
The wall tiles are made of a ceramic mosaic that contains post-industrial recycled material.

Simple lines in the bathroom are always preferable. The bathroom of this home in Belgium features natural elements such as teak. As it is not a local wood, this choice should be supported with a certification on the sustainable origin of the material.

MOD is the new collection by VitrA. Designed by Ross Lovegrove, it consists of a modular system that combines ceramic and bamboo. This shrub, if obtained from sustainable plantation is a good alternative to wood to give a bathroom that desired natural touch.

For those requiring a more luxurious aesthetic, this bathroom is equipped with Agrob Deutsche Buchtal Steinzug ceramic material, manufactured in line with clean technology criteria. Gold colored decorative elements may be a problem, as metallic or gold colors often contain a higher proportion of heavy metals.

As with the living room, kitchen and bedroom, Listone Giordano markets wooden floors finished with natural oils and certified by international bodies such as FSC or PEFC, for flooring, vanity units and floorboards.

Claudio Silvestrin has designed these stone bathtubs and washbasins with simple and natural forms in his collection Le Acque. A return to his roots with extreme elegance.

The ceramic cladding with the Hydrotect finish by DeutscheSteinzug mimics the functions of a lotus flower leaf, through a nanostructure containing hydrophobic waxes, which prevents water and dirt from sticking to it. It is also inspired by the hydrophobic finishes of Teflon pans, which repel water droplets.

Hydrotect consists of titanium dioxide on the surface of the tiles, so that it catalyzes a reaction between light, oxygen and moisture. This photocatalytic process produces activated oxygen, which breaks down microorganisms such as bacteria, fungi and algae.

This type of resistant coating, when applied in the baking process, is not only suitable for walls but also for floors. This film eliminates formaldehyde odors from household furniture, from tobacco, cooking and bathroom odors.

In conventional tiles moisture accumulates in droplets, but In Hydrotect tiles, the water forms a thin film that removes dirt. Moisture and dirt can be removed with a rag. This reduces water used for cleaning which then turns into sewage water and the use of household cleaners for bathrooms that, unless they hold an eco-certificate, often contain products harmful to our health.

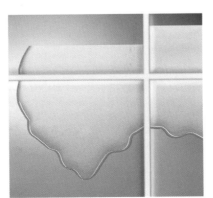
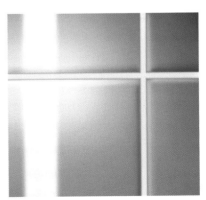

The use of modular furniture saves space, reduces the use of resources and provides clean surfaces in one space, the bathroom, whose principal value is hygiene.

This Zen-ambience bathroom is flooded with natural light thanks to the large central skylight. Natural light is not always possible in a bathroom, especially if you live in an apartment, but when possible, it saves energy and is very beneficial to the owner.

Ceramic mosaics tiles in flooring and walls are frequently used in the bathroom. The vitrified ceramic by Reviglass contains 100% recycled glass in all its products and uses a bonding system with polyurethane cord.

The use of this recycled material reduces energy consumption by 60% during production. According to the manufacturer, the lower the mass of Reviglass mosaic in relation to ceramic results in savings of 25% less energy consumption. Another advantage is that this type of mosaic is laid with a paper panel instead of using glass mesh and PVC points, which generate more polluting waste.

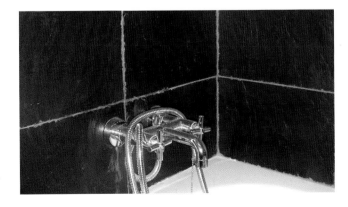

The Barcelona-based company Zicla manufactures synthetic slate with recycled plastic to be used in roofs, façades and bathrooms.

The eTech and eMote series by Dornbracht are automatic electronic faucets that operate with a sensor that promotes hygiene and saves water. These faucets turn on without any contact; the user just has to place their hands under the faucet. This system is often used in commercial and public places, but it is also marketed for home use.

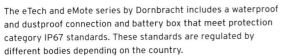

The eTech and eMote series by Dornbracht includes a waterproof and dustproof connection and battery box that meet protection category IP67 standards. These standards are regulated by different bodies depending on the country.

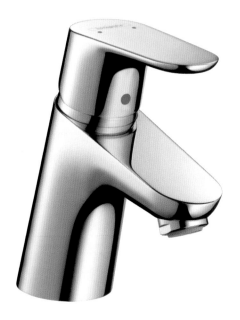

Hansgrohe presents faucets with EcoSmart technology, which limits the flow through special nozzles and the addition of air. This can reduce water consumption from 1.3 to 2.1 gallons per minute without altering the level of comfort. If the output flow is reduced, the flow of water to be heated is also reduced; therefore there are also indirect energy savings.

All mixer faucets by Roca help reduce water consumption by up to 50% thanks to its safety click system. If you require more water, push the faucet upward. They also incorporate a device that helps reduce the water temperature.

Conventional monocontrol faucet

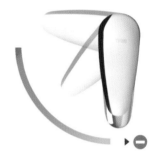

TRES monocontrol faucet

Tres faucets reduce water and energy consumption by 50%. When turned on and the faucet is centrally positioned, cold water runs; hot water can only be used when it is turned to the left.

Another advantage of this system is the easy flow regulation, which is done by simply turning the regulator screw with a coin – below the pipe – and choosing the water pressure you desire. This system received the Environmental Quality Assurance Distinction from the Generalitat de Cataluña, which is an eco-label promoted by the Catalan government.

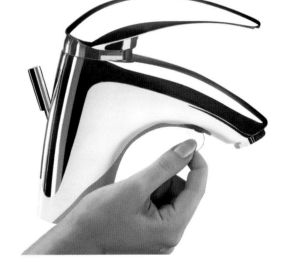

W + W (Washbasin + Watercloset) is a proposal created by the Roca Innovation Lab and designed by Gabriele and Oscar Buratti. This design integrates two essential elements in any bathroom space: the toilet and sink. Combining them in one piece saves space without renouncing an attractive design. Furthermore, by offering a multifunctional product, the resources used and production waste materials are reduced.

This system reduces water consumption by 25%, since the water used in the sink is used to fill the toilet cistern. An automatic cleaning system prevents bacteria forming in the water and odors.

The Kendo-T faucet, with a thermostatic mixer, automatically maintains a constant water temperature that the user selects, regardless of variations in the flow pressure.

The German company Hansgrohe sells hand showers that reduce water consumption. The Raindance collection presents Air Power technology: air bubbles are added to the water passing through all of the showerheads (three liters of air is added to every liter of water). This collection and the model Crometta 85 Green presents EcoSmart technology to limit the flow.

The Irisana IR15 Eco-shower model consists of a shower head with 412 micropores and a filter formed by ceramic particles. Due to a micronization process, the shower head releases large quantities of negative ions (Lenard effect) which are very beneficial to our health, and they cleanse and disinfect the skin without using soap or other products. It reduces the chlorine content and saves up to 65% of water and energy required to heat it. Its price is approximately $75.

The double push button method is widespread in restrooms in developed countries. The building legislation of most countries require their installation in new construction, but there are many existing buildings in which strategies to reduce water consumption undergo other more ingenious solutions.

Water volume reducers are a simple proven and economic device, which helps to save water in toilet cisterns. Some 2 qt. (1.8 l) are saved every time the toilet is flushed. A series of eco-messages have been printed on the packaging of the reducer to increase user awareness. This represents a savings of over 1,000 gal. (3,800 l) of water per person per year.

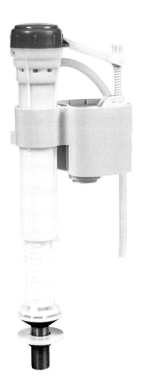

The side and lower filling faucets by Orfesa have a regulating mechanism and the quantity of water in the toilet cistern can be adjusted. They are easy to install.

The use of aerators in bathroom or kitchen faucets can save 50% of water used. Orfesa sells models with a plastic interior that is limescale resistant.
Conventional aerators typically become less efficient over time due to limescale; therefore we recommend that they are regularly cleaned with vinegar.

The use of 2 gal. (9l)/minute stabilizers saves an average of 50–60% of water consumption in comparison to a conventional standard shower. These valves should be placed between the faucet and the flexible hose. This prevents a surge in pressure in the flexible hose and stabilizes the flow, regardless of pressure from the network or by turning on the faucet.

Plants are well-placed in a bathroom as they purify the air. If plants alone do not give you the desired effect, use a dehumidifier such as a fan. We do not recommend the use of PVC curtains as they contain elements with chlorine and phthalates – plasticizers that turn hard plastic into soft plastic.

The ideal relative humidity for both comfort and health is between 45% and 55%, in both summer and winter. According to the principle of condensation, dehumidifiers extract surplus water from the air and store it in a tank. This improves air quality and reduces the likelihood of mold development. They are a wise choice for people with allergies, as they reduce the spread of mites and fungi.

Sancor Industries markets Envirolet composting toilets, which have a very efficient way of saving water. These systems, which do not use water, use hot air to evaporate the water content of waste, a small turbine that extracts excess air and a compost catalyst if you want to accelerate the process. The compost is collected from the bottom compartment about twice a year.

The remote system, which uses just 16 oz. (0.5 l) of water when flushed, can have up to three toilets connected to a single composting unit.

We have discussed good measures for saving water, such as a dual flush system, aerators and showers instead of baths. The most efficient system would be to have two water supplies in the home: a drinking water network, which has to be drinkable, and other network for another uses, fed by a purification and graywater (sink, bathtub, shower and washing machine) reuse system.

The company Alqui-Envas produces tiles and grids made from recycled plastic. This plastic is waterproof and weatherproof (moisture, ice, UV rays) and resistant to corrosion (sea water, acids, oils). These products are lighter than wood and concrete and are certified by the German ecolabel *Der Blaue Engel*.

OUTDOORS

Although this book primarily focuses on the interior of the house, it is also important to consider how to arrange our balcony, patio or small plot of land in accordance with sustainable criteria. These outdoor spaces can be a test lab where materials with a high percentage of recovered components are used, renewable energy devices are installed or, simply, local species are planted in pots or in the garden.

Alqui-Envas also sells recycled plastic pallets for the balcony. This type of product is an alternative to virgin materials, which have a higher embodied energy, as apart from the production process, energy consumed during the extraction must be considered.

The green roof in Kyoto House in Lleida has a mixed use: as an outdoor space, as a purifying agent for nearby rooms and to control the temperature. In summer, the surface of this type of roof does not get as hot and in winter the cold that tends to accumulate on bare indoor terraces is not as severe.

Alqui-Envas also offers the possibility of fencing as well as benches and garden planters for public and private spaces, all made from recycled plastic. These products are certified by the ecolabel *Der Blaue Engel*.

Composite wood or engineered wood is a product that combines the best of wood and the best of plastic resins (polyolefins: polyethylene and polypropylene). eco-Profil composite wood does not require any maintenance, it does not rot or splinter and it is resistant to insects and ultraviolet radiation. Raw materials that use eco-Profil are 95% recovered materials: the wood fiber comes from the sawdust from untreated wood from sawmills and furniture manufacturers in the European market. All suppliers are PEFC certified.

Composite wood can be used in exactly the same way as solid wood; the same technology and types of tools are used. This decking is suitable for outdoor facilities such as decking for gardens and cladding for roofs and walls. The polyolefins used come from post-industrial or post-consumer recycled plastics.

Ecoralia markets the eco-decking Madertec, which is totally rot-proof, does not splinter or decompose, and requires no maintenance. It is highly resistant to moisture and water; therefore it is recommended for use in pool decking. This is a 100% recyclable material.

Gabarró sells Wood Deck for the garden, pool or patio. FSC certified wood is used from different origins: ipe, cumaru, grapia, iroko, masaranduba and flanders wood.

Leopoldo offers this urban garden as an adaptation of the classic garden, a space for growing vegetables, culinary and medicinal plants on the balcony or patio. Its vertical distribution is particularly suitable for small spaces. The L model, which costs about $200, weighs about 7 lb. (3kg) and has a capacity of 21-26 gal. (80-100 l) of substrate.

Another option is to use a galvanized steel cultivation table, measuring 27 x 55 in. (70 x 140 cm), which is easy to assemble (only requires eight screws) and install on patios. These tables have an incorporated drip irrigation system.

The use of composters on balconies in the city is not widespread; they are more often found in a house with a yard. Through the process of composting (aerobic decomposition) the remains of food or garden become humus, a good fertilizer for plants. This process helps us return to the land what it gave us, closing the natural cycle. The image shows a computer graphics model of Composting Combox.

The German company Graf sells composters such as the 70 gal. (265 l) Eco Composter, with a wide opening for easy filling and a removable cover. It is made from recycled polyethylene. The compost is removed through the small front door or by lifting the composter.

The Eco King composter by Graf has a capacity of 79 gal. (300 l) It is made from recycled polyethylene and has a comfortable upper opening. It features an optimized aeration system to speed up the composting process.

Combox is the new model created by the company Compostadores. Its modular design is optimized so that, starting from a basic module of 39 gal. (148 l), more compartments and boxes can be added. Made from 100% post-consumer materials, Combox has a tray base that collects leachates, which facilitates its use on city balconies. The versatility of the Combox is that you can use one or several modules to store yard waste that can later be used as compost.

Vericompost is an aerobic decomposition process involving earthworms. The result of the process is humus, a natural fertilizer rich in nutrients that serves as a fertilizer and soil conditioner. A small Vericompost can be used to convert kitchen waste into high quality fertilizer, especially if space is limited. It is one of the best options if you have a balcony.

Solar-Fizz is a pool shower that can be installed in one minute and only requires a hose connected to the water supply. Its height varies between 4.2 and 7.5 ft. (128 and 229 cm). It has a twin solar collector with and integrated water storage system. Water is heated by the sun and stored about 8 gal. (30 l) per cylinder.

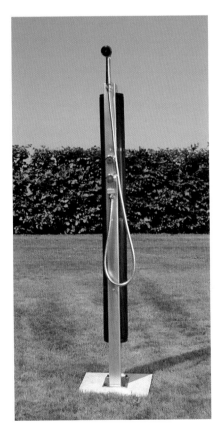

A pool is a wasteful luxury of water that should be avoided. For those unable to abstain, the best option is naturalized pools, as they do not need chemical treatment for the preservation and natural purification of the water, as this takes place beside the open area for bathing and swimming. This system mimics the purification of microorganisms in natural water.

The recommended minimum size for these systems is about 270 ft^2 (25 m^2). Aromatic plants, such as mint water and ornamental aquatic plants are chosen according to the location.

Naturalized pools require minimum maintenance in spring and fall. In summer, however, they require weekly maintenance.

The design of a porch as the entrance to the garden of the house is recommended as it serves as a solar canopy and prevents the overheating of interior spaces. If natural wood is used for the exterior decking, make sure that it is sustainable wood and if possible, locally sourced.

When designing the garden, the investment, hours of work, quality of soil, the light it receives and climate throughout the year must all be taken into account.

The most efficient way to irrigate a garden is through drip irrigation. A moisture sensor detects the soil humidity and irrigates only when necessary. But before deciding on an irrigation option that consumes less water, think about using plants for the garden that do not need to be watered often.

As the image illustrates, in arid and dry climates or low rainfall climates, use local and xerophilous species that are well adapted to drought, as this is the most reasonable option from and environmental viewpoint. Use natural fertilizer.

ACCESSORIES FOR AN ECO HOME

Throughout this book we have been on a fascinating journey of the environmental measures in the design phase of a home, increasing the owner's awareness to renovate the interior of the home. We have looked at technological options with more domestic resources, options involving a higher initial investment and others that only require a change in everyday habits. Next, we list a series of useful home accessories that can be used for various purposes: purifying air in the home for a healthier environment, using photovoltaic solar energy on a small scale, or simply enhancing the use of recycled-content furniture. This all revolves around that great idea that small gestures can have a major impact.

Andrea is a natural air freshener designed by Mathieu Lehanneur and David Edwards, who were inspired by the NASA experiments carried out in the 1980s. The purification process is carried out through the leaves and roots of the plant that is inside the device. The best species to use are *Spathiphyllum*, *Dracaena marginata*, *Chlorophytum comosum* and *Aloe vera*. This new device will answer the question of whether or not it would be better to simply have the plant on its own, rather than producing a new product for this purpose.

Plants around the home, except in the bedroom, are recommended for purifying the air and attenuating noise. The *Live within the system* by Greenmeme is designed with a CAD program and includes plants, drip irrigation, cultivation substrate and lighting points.

Saving water in a priority in many parts of the world. This device was invented by the Australian Ian Alexander in a country in which this element is a precious commodity. It collects the water when we wash dishes or wash vegetables and fruit. Depending on how soiled the water is, it can be used to water plants, or to pour into the toilet after use, or for grooming pets. Its shape is particularly suitable for the kitchen sink.

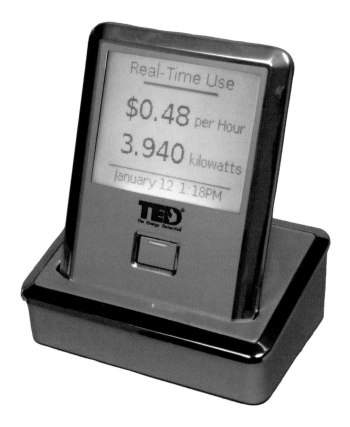

Before applying measures to reduce energy consumption, it would be useful to be aware of the total energy consumption of the home and, in particular, of each appliance.

The wireless electricity monitor TED 5000, by Energy Inc, incorporates a sensor that is installed in the home's breaker panel and transmits data to a remote display, allowing homeowners to see in realtime, the energy that their home is consuming. The data is also transmitted to your home computer, where it can be viewed using the TED footprints software. In addition, the data can be viewed via the internet or any mobile device, and is compatible with Google's PowerMeter.

Homeowners can track their realtime, monthly, average daily, and projected kWh usage, as well as the actual cost based on any number of different rate structures, such as flat, tiered, seasonal or time-of-use.

The TED 5000 comes in a variety of display and purchasing packages, and the cost can range from $200 to $500, depending on the number of electrical panels that are being monitored, and the number of remote display units.

The electricity controller calculates kWh of electricity consumed by household appliances. It has a display and a socket where you connect the appliance. Its price is around $30.

Power strips with a switch are useful to prevent energy consumption of electronic devices in standby mode, which can represent almost 20% of the total home energy consumption. Although for the idle, there will always be a better option. The Belkin Conserve Energy-Saving Surge Protector is remote-control operated, so with a simple button it can completely turn off all electronic equipment connected to one power point. Its price, depending on the dealer, is around $50.

Fundació Terra, through its campaign "Guerrilla Solar," markets the Photonic Kit GS 120. Currently only available in Europe, this solar appliance injects renewable electricity from any socket in the house. It consists of a 120-watt peak power photovoltaic module and a 125-watt inverter connected to the grid, which can generate about 144 kWh of clean electricity annually.

This kit can be self-installed and it only works if there is an electrical network. The kit disconnects when the power goes out, as would a conventional consumer appliance, not to create an island effect. It should be positioned facing south.

The kit is designed to self-consume the electricity generated and save electricity and CO_2 emissions. Ideal for patios (check local town planning regulations) or private yards. The price of the kit is around $1,000.

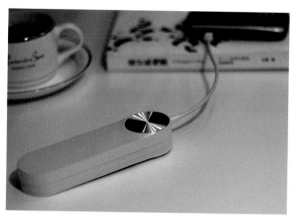

Sunny Flower, designed by Fandi Meng, is a solar charger. It works like a flower, the petals open, absorbing the sun's rays. If you can achieve the correct orientation, it will stick to glass. It allows you to charge a mobile phone or MP3 player.

HYmini is a charger for mobile phones, MP3 players (depending on model), batteries if you have a USB adapter and PDAs. The device has a USB port with various connectors that can charge electronic devices, up to 5 V. A photovoltaic solar minipanel can be added to the device.

The charger, which weighs 3.17 oz. is designed for use while cycling or running, however it can also be installed on a patio or balcony (the atmosphere must be dry and a wind speed over 9 mph or 15 km/h). The price is about $75.

For a house with the correct solar latitudes, solar cookers are highly recommended because they do not pollute or emit fumes. The company alSol markets the model alSol 1.4, which costs approximately $325. It generates heat energy similar to that of a 600W electric burner. These cookers work by concentrating the heat that the aluminum surfaces reflect. This heat is distributed around the cooking pot (which should be black to better absorb the heat) and cooks all types of food.

Two people can set up the alSol 1.4 cooker in under two hours. Ideally it is installed on a balcony or in the yard.

Examples of cooking times: coffee for six people, 10 min; 11 lb. (5 kg) of oven roasted potatoes, 60 min; a sponge cake, 45 min; and paella for 10 people, 90 min. Its expected useful life is 20-30 years. It weighs about 20 lb. (9 kg) and its dimensions are 41 x 16 x 2 in. (10 x 41 x 5 cm).

In developing countries its use prevents deforestation as wood is not used for cooking.

Lasentiu manufactures the ZIG-ZAG bottle rack. It measures 22 x 12 x 3 in. (56 x 30 x 8 cm), and it is made of Syntrewood, a plastic material made from domestic packaging. It is water-repellent and contains no glues or adhesives.

The basic unit formed by four V-shaped sections can be stacked one on top of the other or they can be placed horizontally.

The Barcelona-based company nanimarquina was founded in 1995 with the main objective of wiping out child labor in the rug manufacturing industry in countries like India, Pakistan and Nepal. The company cooperates with the international organization Care & Fair.

This rug from the Bicicleta collection has emerged from research on the possibility of using recycled rubber to create new textures. The rugs are produced in India with locally sourced bicycle tubes. They are handcut into strips and hand-woven on looms.

The Ambara Long (long pile) and Ambara Short (short pile) collection by Francisco Cumella consists of 80 x 120 in. (200 x 300 cm) handmade rugs made from strong, durable cannabis hemp fiber. The dye is natural and they cost from $1,000 to $1,500 each.

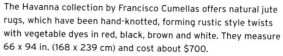

The Havanna collection by Francisco Cumellas offers natural jute rugs, which have been hand-knotted, forming rustic style twists with vegetable dyes in red, black, brown and white. They measure 66 x 94 in. (168 x 239 cm) and cost about $700.

If you want to buy a rug that respects the environment, the following guidelines should be followed: It should be handmade from natural raw materials (wool, silk, cotton, jute, hemp or any recycled material) and with natural dyes that do not contain adhesives, which can give off VOCs such as benzene, toluene or formaldehyde. Unfortunately, wool carpets usually contain the pesticide found in mothballs.

We recommend that you periodically clean the carpet to avoid allergic reactions, breathing problems and the appearance of mold and mites. Every three to five years it should be hand-washed by a professional.

MAINTENANCE

In this book we have covered measures, means and materials necessary to make the home healthy. In this section we will discuss solutions for the cleaning, conservation and periodic maintenance of the home. Durable materials are more sustainable as they represent a saving in resources.

When painting interior and exterior walls, handrails or rails, or interior or exterior woodwork, the best option is to use glue, lime or natural resin-based paints, which are colored with the addition of soil, gypsum and plant pigments. Vegetable-based paints are inexpensive, and easy to use. They are in places that are not exposed to moisture due to their poor water resistance.

Lime-based paints are breathable and have an antiseptic and antifungal function, thanks to the antibacterial properties of lime. They can be used outdoors and in wet environments or in environments with extreme temperature changes.

Natural resin-based paints are usually more expensive and are not available in as many colors.

Paints, varnishes and solvents that are not natural release volatile organic compounds (VOCs), which increase the ozone levels in the troposphere – the atmospheric layer closest to Earth.

Until the 1940s, pigments and solvents were natural. They were replaced by synthetic dyes, chemicals and petroleum products. The best paints are linseed or soy oil-based, or those made from coniferous resins, such as larch and Scots pine, or beeswax, rosin, starch, wax, shellac, casein, in particular for materials such as wood.

Plaster is used to protect and smooth the surface of interior and exterior walls; it also acts as a base on which to apply paint or other coatings. It can be used as an acoustic insulator, it makes wood more fire resistant and it absorbs surface moisture on interior walls.

The best plasters are made up of a lime and sand-based mortar. Plasters from synthetic resins should not be used, and the use of cement should be limited.

Marble is a porous material; therefore stains should be cleaned immediately with soap and water. Rust and grease stains are removed by scrubbing with paper towel dampened with acetone (wear gloves when using this substance as it is caustic).
As a precaution, to care for the marble rub it with the inside of an apple skin. To remove tea and coffee stains use three tablespoons of borax in a glass of water. Other stains can be treated with a mixture of water and hydrogen peroxide. Avoid contact with acidic products.

To improve the properties of wood, accentuating the grain and color, you can apply natural oils such as linseed, natural resins or beeswax.
Unnatural and petrochemical treatments can be harmful to your health. To remove parasites simply use substances extracted from the bark and leaves of the tree itself.

Livos sells natural oils and waxes used to treat wood. These plant or mineral-based treatments guarantee the maintenance of this material without emitting formaldehyde.
To preserve parquet it should not be exposed to sunlight for prolonged periods of time. During the summer or when the heating is on high, containers of water can be placed around the house to maintain the humidity. For stains, lightly moisten and dry immediately.
It is also useful to put plastic bases or protectors on the legs of furniture, especially if the furniture is heavy, to prevent scratching.

Bamboo floorboards can be cleaned with a mop or a dust rag. Occasionally, use a cloth soaked in warm water with a dash of vinegar. Compared with wood, bamboo is extremely water resistant, making it suitable for bathrooms and kitchens.

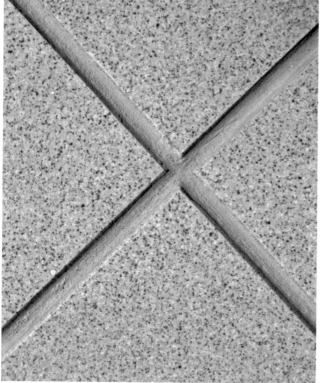

In the case of ceramic tiles or stone, we recommend:
- Regularly scrub the floors with vinegar and water.
- Products containing polish are not recommended, because they can attract dust.
- On unglazed ceramics and interiors, you can apply natural waxes and polishes to achieve shine and improve wear.
- Every five years check the joints between tiles.

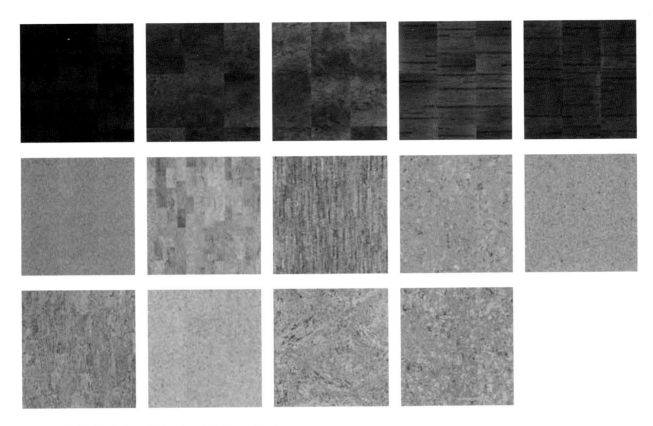

This palette of colors belongs to the almost 50 types of cork for the Corkcomfort by Wicanders line of flooring, marketed by Amorim. To clean a cork surface, use boiling water and a little salt. If it is stained, add vinegar or eucalyptus oil, which will disinfect the surface and leave a pleasant fragrance.

Removing grit from the surface of linoleum can prevent damage by abrasion. According to the manufacturer, if not excessively dirty, we recommend that you clean when wet by adding a neutral or alcohol-based detergent to the water. If the dirt is more persistent, you should use an automatic disk polishing machine or a cleaning agent with a pH that does not exceed 10. The natural alternative is to clean the surface with vinegar and hot water with a neutral liquid soap.

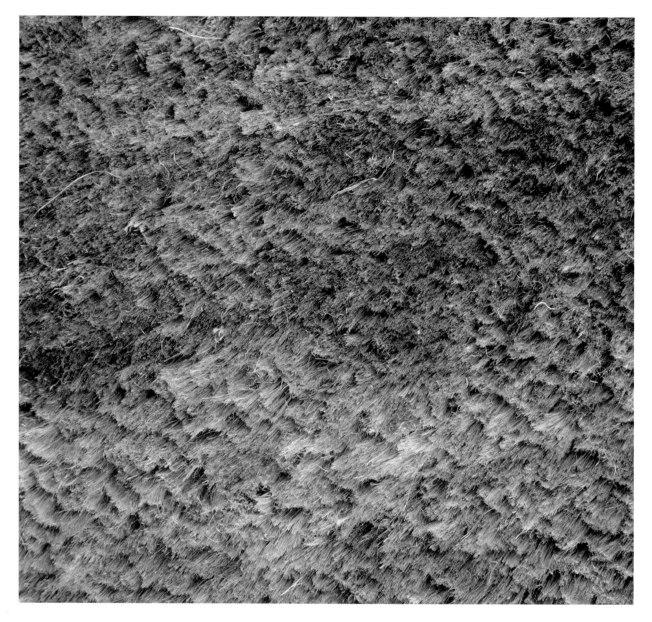

Rugs are cleaned with water and a little vinegar, which helps fix the color. Shake a bit of baking soda on the stain before vacuuming. If it smells bad, sprinkle baking soda over the entire surface, let it settle for 30 minutes and then vacuum.

Before using a chemical air freshener, think about whether or not a few plants spread throughout the house could achieve the same effect. Plants produce oxygen, absorb carbon dioxide, regulate moisture and the temperature, reduce noise and retain dust. The most desirable for their purifying effect are:

- Dracena (*Dracaena marginata*): filters formaldehyde and xylene
- Palmera (*Chamaedorea*): filters formaldehyde and trichlorethylene
- Cactus (*Cereus peruvianus*): compensates for the radiation from computers
- Hiedra (*Hedera helix*): filters benzene and formaldehyde. Ideal beside computers, fax machines and other appliances
- Cinta (*Chlorophytum comosum Variegatum*): filters 96% of carbon monoxide, xylene and formaldehyde.

CLEANING THE HOME USING FEWER CHEMICAL PRODUCTS

Most conventional detergents include chemicals or toxic substances harmful to humans and the environment. Developing or enhancing the quality of life, as we understand in the West, often equates to the convenience that many consumer products offer us, therefore natural treatments have been neglected in many homes for individual solutions for each type of surface or chemical product that promise a lot more than they can offer. We do household chores without worrying about the contaminants in the water from our pipes.

Below are some tips on how to clean just as our grandmothers did, which had much less impact on the environment than current trends in household cleaning.

General

- Detergents should be 100% biodegradable
- Use vinegar as a disinfectant, polish and softener
- Use lemon juice to polish metals
- Be suspicious of scents
- Avoid the use of aerosols
- Always read the label: the more hazard symbols, the worse it is for our health and the environment
- Use washable cotton rags instead of disposable rags or paper towels

CLEANING AGENT	HOW TO MAKE	AVOID
Disinfectant	A quarter cup of borax dissolved in a pint of hot water. Add eucalyptus oil.	
Stain remover	- Option A: cold water and vinegar in equal parts, with essential oils (rosemary or lavender). For blood stains, use sodium cardonate instead of vinegar. - Option B: alcohol or acetone.	Chlorinated solvents.
All-in-one cleaner	Dissolve two tablespoons of borax and one tablespoon of soap in a quart of water.	
Floor cleaner	Solution of water with lemon, vinegar, borax or baking soda.	
Drain cleaner	Recommendation: Pour baking soda and vinegar down the drain. Leave for several minutes. Rinse with boiling water.	Chlorinated products and acid products.
Air-freshener	- Option A: water simmered with cinnamon sticks, cloves or lavender. - Option B: bowl of dried flowers, petals or spices.	

CLEANING POINT	TIP	AVOID
Laundry	- Use an eco washing ball that cleans by producing ionized oxygen without detergent or fabric softeners. - If the eco ball is not enough, use eco-label detergent. - Dry in the sun whenever possible.	Tumble dryers, if the weather where you live permits it
Furniture	Rub with a solution of lemon juice (1 tsp./5 ml) and olive oil (1 cup/250 ml).	
Bathroom (bathtub, tiles and sink	Rub with a sponge with baking soda, borax or salt.	
Faucets with limescale	Scrape the limescale, apply boiling water with salt and vinegar. For tough limescale, use undiluted vinegar.	
Mirrors	A spray solution of warm water and alcohol.	
Glass windows	Clean with soap and water. Then dry with newspaper dampened with methylated spirits.	
Slime or mold	For steel, use vinegar. For brass and copper, use a solution of vinegar and salt.	
Rugs and carpeting	Clean with water and a little vinegar. If stained, let baking soda settle for 30 minutes and vacuum.	
Marble	Clean with mild soap and water.	Contact with acids
Wood parquet	Clean with a damp cloth and dry immediately.	- Hours of direct sunlight - Contact with moisture
Bamboo parquet	Clean the dust with a rag. In case of stains, moisten a cloth with warm water and vinegar.	
Stone flooring	Clean with water and vinegar.	
Ceramic flooring	Clean with water and vinegar.	
Cork flooring	Clean with boiling water with a little salt. If stained, use vinegar.	
Linoleum	Solution of hot water with vinegar and neutral liquid soap.	Soak the surface
Aluminum	Clean with lemon juice.	
Steel	- Option A: to remove rust, use hot oil. - Option B: rub the rust stain with an onion cut in half and then wash with oil.	
Copper	Rub with an onion cut in half, then with half a lemon sprinkled with salt.	
Brass	Clean with lemon juice. If very dirty, use a rag with vinegar and salt.	

DIRECTORY OF ARCHITECTS AND DESIGNERS

1+2 Architecture
Hobart, TAS, Australia
www.1plus2architecture.com

24H architecture
Rotterdam, The Netherlands
www.24h.eu

Agence Coste Architectures
Houdan, France
Montpellier, France
www.coste.fr

Altius Architecture
Toronto (Ontario, Canada)
www.altius.net

Álvaro Ramírez B
Santiago de Chile, Chile
www.ramirez-moletto.cl

Arkin Tilt Architects
Berkeley, CA, United States
www.arkintilt.com

Atelier Werner Schmidt
Trun, Switzerland
www.atelierwernerschmidt.ch

Carter + Burton
Berryville, VA, United States
www.carterburton.com

Clarisa Elton
Santiago de Chile, Chile
http://clarisaeltonarquitecta.blogspot.com/

Claudio Silvestrin Architects
London, United Kingdom
www.claudiosilvestrin.com

Dan Rockhill/Rockhill + Associates
Lecompton (Kansas, United States)
www.rockhillandassociates.com

Dietrich Schwarz
Domat/Ems, Switzerland
www.schwarz-architektur.ch

Fandi Meng
Shenzhen, China
www.fandimeng.com

FAR Frohn & Rojas
Cologne, Germany
Santiago de Chile, Chile
Los Angeles, CA, United States
www.f-a-r.net

Friman, Laaksonen Arkkitehdit Oy
Helsinki, Finland
www.fl-a.fi

Gernot Minke
Kassel, Germany
www.gernotminke.de

Giovanni D'Ambrosio
Rome, Italy
www.giovannidambrosio.com

GLASSX AG
Zurich, Switzerland
www.glassx.ch

Greenmeme
Los Angeles, CA, United States
www.greenmeme.com

Green Fortune AB
Estocolmo, Sweden
www.greenfortune.com

Greta Pasquini
Paris, France
g.doron@free.fr

John Friedman Alice Kimm Architects
Los Angeles, CA, United States
www.jfak.net

Kieran Timberlake Associates
Philadelphia, PA, United States
www.kierantimberlake.com

Kirkland Fraser Moor
Aldbury, United Kingdom
www.k-f-m.com

Luca Lancini
Barcelona, Spain
www.lucalancini.com

Marcio Kogan
São Paulo, Brazil
www.marciokogan.com.br

Mario Alberto Tapia Retama
Ciudad Obregon, Mexico
www.reciclajeecologico.ning.com

Markus Wespi Jérôme di Meuron Architekten
Caviano, Switzerland
www.wespidemeuron.ch

Martin Liefhebber/Breathe Architects
Toronto, Ontario, Canada
www.breathebyassociation.com

Michelle Kaufmann Studio
United States
www.michellekaufmann.com

MITHUN
Seattle, WA, United States
www.mithun.com

Nicola Tremacoldi/NM Arquitecte
Sant Cugat del Vallès, Spain
nicmanto@coac.net

Obie G. Bowman
Healdsburg, CA, United States
www.obiebowman.com

Olgga Architects
Paris, France
www.olgga.fr

OMD - a Jennifer Siegal company
Venice, CA, United States
www.designmobile.com

Petz Scholtus/Pöko Design
Barcelona, Spain
www.pokodesign.com
www.r3project.blogspot.com

Pich-Aguilera Arquitectos
Barcelona, Spain
www.picharchitects.com

Pugh + Scarpa
Santa Monica, CA, United States
www.pugh-scarpa.com

Rongen Architekten
Wassenberg, Germany
Erfurt, Germany
Düren, Germany
Chengdu, China
www.rongen-architekten.de

Sambuichi Architects
Hiroshima, Japan
samb@d2.dion.ne.jp

Siegel & Strain Architects
Emeryville, CA, United States
www.siegelstrain.com

Simon Swaney
Sydney, NSW, Australia

System Architects
New York, NY, United States
www.systemarchitects.com

Three House Company
Wollaston, United Kingdom
www.treehousecompany.com

Tonkin-Zulaikha-Greer Architects
Surry Hills, NSW, Australia
www.tzg.com.au

Verdickt & Verdickt architects
Antwerp, Belgium
www.verdicktenverdickt.be

Vicens + Ramos
Madrid, Spain
www.vicens-ramos.com

DIRECTORY OF MANUFACTURERS AND ASSOCIATIONS

ALQUI-ENVAS
www.alquienvas.com

alSol Technologías solares
www.alsol.es

Amorim Cork America
www.amorimcorkamerica.com

Andrea/LaboGroup
See www.andreaair.com for U.S.
availability.

Armstrong
www.armstrong.com

Artquitect
www.artquitect.net

Auro Pflanzenchemie AG
www.auro.de

Bioklima Nature
www.bioklimanature.com

Bioteich/J.N. Jardins Naturels
www.bioteich.fr
Canada: mhudon@val-mar.ca

Belkin International Inc.
www.belkin.com

Cannabric
www.cannabric.com

Chromagen
c/o AO Smith
www.chromagen.biz

Compostadores
www.compostadores.com

Cosentino
www.cosentinonorthamerica.com

De'Longhi
www.delonghiusa.com

Delta Faucets
www.deltafaucet.com

Deutsche Steinzeug America, Inc.
www.deutsche-steinzeug.de

Dornbracht Americas Inc.
www.dornbracht.com

Ecoralia
www.ecoralia.com

EdilKamin
www.edilkamin.com

Efergy/Efermeter
www.efergy.com

Energy Inc.
www.theenergydetective.com

Envirolet/Sancor Industries Ltd.
www.sancorindustries.com

Forest Stewardship Council
www.fsc.org

Francisco Cumellas
www.franciscocumellas.es

Fratelli Spinelli
www.fratellispinelli.it

Hansgrohe Inc.
www.hansgrohe-usa.com

Hispalyt – Spanish Association of
Manufacturers of Clay Bricks and
Roofing Tiles
www.hispalyt.es

Home Energy International
www.homeenergyamericas.com

Hughie Products Pty. Ltd.
www.hughie.com.au

HYmini/MINIWIZ
www.hymini.com

Jaga Inc.
www.jaga-usa.com
www.jaga-canada.com

Lasentiu
www.lasentiu.com

Leopoldo Group Design
www.leopoldobcn.com

Listone Giordanó/Margaritelli Ibérica
www.listonegiordano.com

Keim Mineral Coatings of America
www.keim.com

nanimarquina
www.nanimarquina.com

Nousol Nuevas Energías
www.nousol.com

Orfessa
www.orfesa.net

Osram Sylvania
www.sylvania.com

Otto GRAF
www.graf-water.com

P3 International
www.p3international.com

Piera Ecocerámica
www.pieraecoceramica.com

Porcelanosa
www.porcelanosa-usa.com

Reviglass
www.reviglass.es

Roca Sanitario
www.roca.com

Runtal Radiators
www.runtalnorthamerica.com

Saunier Duval
www.saunierduval.es

Solar-Fizz
www.gartendusche.com

Solicima
www.soliclima.es

Sonkyo Energy
www.sonkyoenergy.com

Tres
www.tresgriferia.com

Tuka Bamboo
www.tukabambu.com

Turbo aire/Seabreeze Electric Corp.
www.seabreeze.ca

Verde 360
www.verde360.net

Weole Energy
www.weole-energy.com

Wicanders
www.wicanders.com

Zicla
www.zicla.com

ONLINE RESOURCES

www.leed-homes.net
www.greenhomebuilding.com
www.dsireusa.org
www.eeba.org
www.greenroofs.org
www.terra.org
www.buildingtradesdir.com
www.coolroofs.org
www.greenbuildingadvisor.com
www.buildinggreen.com
www.cagbc.org
www.treehugger.com
eartheasy.com
www.enerworks.com
www.aceee.org
www.enviroharvest.ca
www.gaiam.com
www.watermiser.com
www.cgmi.org
www.houseneeds.com
www.awea.org